Writing about Science

WRITING ABOUT SCIENCE

Second Edition

Edited by

ELIZABETH C. BOWEN

BEVERLY E. SCHNELLER
MILLERSVILLE UNIVERSITY OF PENNSYLVANIA

New York   Oxford
OXFORD UNIVERSITY PRESS
1991

Oxford University Press

Oxford   New York   Toronto
Delhi   Bombay   Calcutta   Madras   Karachi
Petaling Jaya   Singapore   Hong Kong   Tokyo
Nairobi   Dar es Salaam   Cape Town
Melbourne   Auckland

and associated companies in
Berlin   Ibadan

Published by Oxford University Press, Inc.,
200 Madison Avenue, New York, New York 10016

Oxford is a registered trademark of Oxford University Press

Library of Congress Cataloging-in-Publication Data

Writing about science /
edited by Elizabeth C. Bowen, Beverly E. Schneller. 2nd ed.
p. cm.
ISBN 0-19-506274-4
1. Science. I. Bowen, Elizabeth C., 1944–
II. Schneller, Beverly E.
Q171.W876 1991 808'.0665—dc20   90-33943

The following pages are regarded as an extension of the copyright page.

9 8 7 6 5 4 3 2 1
Printed in the United States of America
on acid-free paper

To Our Parents

Acknowledgments

The editors wish to thank Philip Leininger and William Sisler of Oxford University Press for their guidance and support throughout this project. Gratitude is also owed to the library staffs at Marist College and Millersville University for their assistance in locating source materials for this book, and to Gregory Jerozal for valuable suggestions.

Contents

Writing for Professional Audiences, 179

Rhetorical Table of Contents

Argument and Persuasion

VARIETIES OF SOURCE AND METHOD

Introduction

Imagine a scientist. One in a movie, or maybe the research physician in the aspirin ads. Or an astronomer, or an industrial chemist. Or, perhaps, the kind of scientist you might someday become.

Now, a question: does this imaginary scientist write well? Unfortunately, probably not too well at all. In most people's minds, and certainly in many humanists', scientists are notoriously bad writers. They prefer jargon to clear expressions, sacrifice the beauties of the active voice on the altar of scientific impersonality, and, even worse, they are usually blind to the imaginative side of the ideas they work with, the "big picture." Writers, on the other hand, are passionate, intuitive, and even though sometimes less precise, they have a sense of mystery, and they love graceful, clear prose. Writers labor at the forge of language to create the consciousness of the race; scientists wear white lab coats and record even columns of numbers on graph paper.

But in reality, many great scientists have also been great scientific writers. This anthology aims to demonstrate the range, the rhetorical complexity, and the sheer excitement of scientific writing; to show that even if it is true that some scientists do not write well, it *need not* be true, for it is not true of these. Some in this anthology come from the British tradition of popular science writing—Michael Faraday, Julian Huxley, J. B. S. Haldane. Others are American—Stephen J. Gould, Rachel Carson, Lewis Thomas, James Watson. Some excel at explaining science to the layman—witness the prolific Isaac Asimov. Others—Gerald Holton, Garrett Hardin—wrestle commendably with the complex demands of communicating to pre-professional or professional audiences. Certainly students of writing—science students to whom this anthology is addressed—can profit from immersing themselves in the words of these scientists.

We hope that this anthology will play some role in challenging the cultural stereotype we describe above and perhaps in lessening whatever truth it has. We know that science students are ill-served by traditional freshman English anthologies and that writing can be taught equally well when the subjects of composition are scientific ones. Our aim is to present models of good scientific writing and, through the study questions, to challenge both student and teacher to consider them as writing, not solely for their intellectual content but for their rhetoric and style.

The first section of this anthology contains popular science writing. Depending upon the audience—whether the generally educated layperson or the whole reading public—the authors here assume some degree of scientific knowledge, and try to link their subject with the world of human experience. Thus, Lewis Thomas's microscopic events, his plankton and mitochondria, challenge some of our cultural evasions, because here death is all around us. And Rachel Carson, the herald of the age of ecological awareness: *Silent Spring* was her powerful statement of the hidden effects of human domination of nature. By opening this section with writing addressed to children (Michael Faraday's "The Chemical History of a Candle"), we do not mean to imply that such writing is easy. On the contrary, writing for children forces the writer to master the subject completely. It can be a humbling experience.

The second section of the book presents various examples of professional and pre-professional communication: scientists talking to other scientists. Almira Phelps, in addressing college students, summarizes the current state of knowledge about electricity in such a way that readers are prepared to assimilate new developments as they continue in training. Gerald Holton, writing for professional physicists, succinctly presents a world view which antedates the one in which they themselves frame speculations. But James Watson and Francis Crick, announcing a structure for the DNA molecule, addressed a small community of researchers working on the same problem.

In selecting material, we have tried to represent classics as well as contemporaries in scientific writing and as many fields of the physical and biological sciences as is possible in a book of this size. At the same time, we have included a number of selections on one subject, DNA, so that a reader may follow the topic through changes of audience and purpose. We have limited ourselves to essays written originally in English, even though this decision has meant excluding many scientists who are masterful writers. But we have included two selections originally delivered as lectures, Michael Faraday's "The Chemical History of a Candle" and one of Richard P. Feynman's California Institute of Technology lectures on physics, "Physics: 1920 to Today."

This anthology emphasizes the history of science. Whatever the level of its audience, any subject for scientific writing is only enhanced by a history of developments in the field. George and Muriel Beadle on Mendel, Stephen Toulmin and June Goodfield writing on Buffon are masters of the history of science: reading their work is training in how to conceptualize the development of a scientific problem.

Expository essays can be as rewarding to the reader, and as demanding of the writer, as the greatest belles lettres, essay or story. We believe the scientific writing in this anthology demonstrates the variety and complexity of subject matter, the rhetorical sophistication, and the skillful adaptation of tone to audience, of the genre. But we also hope that we show that scientific writing, like all great writing, is self-referential. By the very act of telling us about the world, these authors say more than perhaps even they are aware they are saying about themselves. To the Beadles, scientific discovery is a blend of talent and chance; to James Watson, it is an exhilarating race to solve a puzzle; to Stephen Gould, science is man trying to understand forces that are much, much greater than he is. If we listen carefully, then, these writers will tell us a very great deal about why they are also scientists.

Writing about Science

Writing for popular audiences

Popularized science has been called "science without tears." One estimate haughtily suggests that only .01% of all scientific information can be communicated to the public. But even though only a small part of science is its subject, the audience for popularized science is large. Indeed, surveys show that many consumers of popularized science are involved professionally in science and are simply curious about fields other than their own.

A clue to the riddle of its proliferation is that science popularization varies widely, according to the interests and backgrounds of its intended audience. There are many kinds of audiences, and a good writer will engage his on the level of its interests. This is the first principle of communication, one which the readings in this section of the anthology are chosen to demonstrate.

How do these authors address their audiences? Michael Faraday's audience of school-children had given up part of the Christmas holidays to come to hear his lecture "The Chemical History of a Candle," and he responded to this by putting on a show, amazing them with the sheer variety of candles, nearly creating an explosion to illustrate how fumes remain after a candle flame is put out. To see capillary action at work in a blue liquid traveling up a column of salt was equally exciting. Science might have seemed a bit like magic to the children who filed out of Faraday's lecture hall.

Popularization addressed to adults is most successful when it brings science into the sphere of adult experience. George Gamow illustrates the properties of curved space by examples so basic that we cannot help but understand him. He reflects upon the place of man in nature and he develops his argument by basing it in the universal language of common experience.

Lewis Thomas and Howard Ensign Evans, on the other hand, show us how understanding science can enrich our experience of everyday life. Once we have read about trees filled with fireflies flashing in unison, that image will surface to memory when on summer evenings we see fireflies light against the dark air. And before reading Lewis Thomas, we might never have known that, like all other animals, we communicate by "vibes," chemicals we leave around us wherever we go.

But with relativity theory and quantum mechanics, we seem to have a science that is outside experience and even alien to common sense. Actually, while it is true that these two fields have certain elements which are intellectually disturbing, it is not true that they contradict the logic of common sense. However, both their premises and some of their conclusions are alien to common experience. For Bertrand Russell, space-time still somehow remains outside the common-sense conceptual world in which we carry on our everyday lives.

Popularized science has had a recent resurgence in respectability. A resurgence, because it has, in the past, been more respectable than it later came to be. Early in the Renaissance, Kepler and Galileo both wrote engaging accounts of their work which they addressed to monarchs. In the seventeenth and eighteenth centuries, the literate bourgeoisie, as well as the nobility, took an interest in science; in fact, there was a time when it was the mark of a gentleman to set aside a room as a laboratory. This amateurism in science soon declined, but it remained commonplace and even fashionable for scientists in the nineteenth century to go out of their way to explain science to the layman. Thomas Edison held a famous "electric breakfast" for which the food was cooked by electricity. Motives for the breakfast were economic: the proceeds from it funded further research. But in the half century before World War I, the increasing professionalization of science led to the proliferation of jargon. Scientists spoke to their own peers, respected only pure research, and looked down on popularization as "vulgarization." In the decades after World War II, even teaching undergraduates was barely tolerated.

But now, there is a greater educated public, and it demands popularization of science. Yet also now, science is becoming increasingly unintelligible to the ordinary man. No special education was necessary to understand Torricelli's experiments, but relativity demands a solid mathematical background. The split between the two cultures widens, and only the most determined keep up with both. A gulf of language and experience seems to separate the scientist in his specialty from the larger community.

Now the popularizer has a new task: to humanize science. Science is a human endeavor; hence James Watson publishes his autobiographical account of unlocking DNA. The popularizer wants to show that behind every discovery there is a human being who lives in the same world we live in. At the same time, the writer cannot convey the impression that science is easy or ordinary.

To explain the popularity of scientific writing we would probably have to call upon an essential fact of human nature. People simply are curious about the world they live in. Perhaps they believe that if they understand it, they control it; perhaps they enjoy understanding it just as they would enjoy seeing a good film or listening to a symphony orchestra. Whatever the reason, in an era of government funded research, a scientifically educated public cannot help but benefit science as well. And popularized science is not without professional value. One of the earliest books to awaken the mind of the young Charles Darwin to the natural world was Robert Sears's compendium, *The Wonders of the World in Nature, Art, and Mind, comprising a complete Library of Useful and Entertaining Knowledge.* Complete with several hundred engravings, its author claimed that it was "without rival in this bookmaking age"! Perhaps his words—"useful and entertaining"—are words to keep in mind when writing popularization of this kind.

MICHAEL FARADAY

The Chemical History of a Candle

MICHAEL FARADAY (1791–1867) was born into the English working class. Hence his formal education was minimal, but he made up for the lack of it by voracious reading. Apprenticed at fourteen to a bookseller, he came upon the article "Electricity" in the *Encyclopaedia Brittanica* and from that moment sought to devote his life to science. Through a fortunate chance, Faraday became Sir Humphrey Davy's assistant at a time when Davy was working on an atomic theory that posited a mathematical point with shells of attractive and repulsive force surrounding it. This became the basis for Faraday's later work on electricity. In 1831, winding an iron ring with two coils (one connected to a voltaic battery, the other leading away to convert the vibrations into a current), he discovered electromagnetic induction. He made important discoveries in magnetism and electrolysis and is considered the inventor of field theory. One of the great English popularizers of science, his Friday Evening Discourses at the Royal Institution became a well-loved tradition. "The Chemical History of a Candle," one of those discourses, is still widely reprinted.

Faraday's expository style is to attack a problem directly, with no flourishes, to expound the problem at hand, to narrate his experiments (including those that failed), leading the reader along to his conclusions. Simple, rustic, dignified.

A Candle: The Flame—Its Sources— Structure—Mobility—Brightness

I purpose, in return for the honor you do us by coming to see what are our proceedings here, to bring before you, in the course of these lectures, the Chemical History of a Candle. I have taken this subject on a former occasion, and, were it left to my own will, I should prefer to repeat it almost every year, so abundant is the interest that attaches itself to the subject, so wonderful are the varieties of outlet which it offers into the various departments of philosophy. There is not a law under which any part of this universe is governed which does not come into play and is touched upon in these phenomena. There is no better, there is no more open door by which you can enter into the study of natural phi-

7

deposited there, not in a careless way, but very beautifully in the very midst of the centre of action, which takes place around it. Now I am going to give you one or two instances of capillary attraction. It is that kind of action or attraction which makes two things that do not dissolve in each other still hold together. When you wash your hands, you wet them thoroughly; you take a little soap to make the adhesion better, and you find your hand remains wet. This is by that kind of attraction of which I am about to speak. And, what is more, if your hands are not soiled (as they almost always are by the usages of life), if you put your finger into a little warm water, the water will creep a little way up the finger, though you may not stop to examine it. I have here a substance which is rather porous—a column of salt—and I will pour into the plate at the bottom, not water, as it appears, but a saturated solution of salt which can not absorb more, so that the action which you see will not be due to its dissolving any thing. We may consider the plate to be the candle, and the salt the wick, and this solution the melted tallow. (I have colored the fluid, that you may see the action better.) You observe that, now I pour in the fluid, it rises and gradually creeps up the salt higher and higher; and provided the column does not tumble over, it will go to the top. If this blue solution were combustible, and we were to place a wick at the top of the salt, it would burn as it entered into the wick. It is a most curious thing to see this kind of action taking place, and to observe how singular some of the circumstances are about it. When you wash your hands, you take a towel to wipe off the water; and it is by that kind of wetting, or that kind of attraction which makes the towel become wet with water, that the wick is made wet with the tallow. I have known some careless boys and girls (indeed, I have known it happen to careful people as well) who, having washed their hands and wiped them with a towel, have thrown the towel over the side of the basin, and before long it has drawn all the water out of the basin and conveyed it to the floor, because it happened to be thrown over the side in such a way as to serve the purpose of a siphon. That you may the better see the way in which the substances act one upon another, I have here a vessel made of wire gauze filled with water, and you may compare it in its action to the cotton in one respect, or to a piece of calico in the other. In fact, wicks are sometimes made of a kind of wire gauze. You will observe that this vessel is a porous thing; for if I pour a little water on to the top, it will run out at the bottom. You would be puzzled for a good while if I asked you what the state of this vessel is, what is inside it, and why it is there? The vessel is full of water, and yet you see the water goes in and runs out as if it were empty. In order to prove this to you I have only to empty it. The reason is this: the wire, being once wetted, remains wet; the meshes are so small that the fluid is attracted so strongly from the one side to the other as to remain in the vessel, although it is porous. In like manner, the particles of melted tallow ascend the cotton and get to the top; other particles then follow by their mutual attraction for each other, and as they reach the flame they are gradually burned.

Here is another application of the same principle. You see this bit of cane. I have seen boys about the streets, who are very anxious to appear like men, take a piece of cane, and light it, and smoke it, as an imitation of a cigar. They are enabled to do so by the permeability of the cane in one direction, and by its capillarity. If I place this piece of cane on a plate containing some camphene (which is very much like paraffine in its general character), exactly in the same manner as the blue fluid rose through the salt will this fluid rise through the piece of cane. There being no pores at the side, the fluid can not go in that direction, but must pass through its length. Already the fluid is at the top of the cane; now I can light it and make it serve as a candle. The fluid has risen by the capillary attraction of the piece of cane, just as it does through the cotton in the candle.

Now the only reason why the candle does not burn all down the side of the wick is that the melted tallow extinguishes the flame. You know that a candle, if turned upside down, so as to allow the fuel to run upon the wick, will be put out. The reason is, that the flame has not had time to make the fuel hot enough to burn, as it does above, where it is carried in small quantities into the wick, and has all the effect of the heat exercised upon it.

There is another condition which you must learn as regards the candle, without which you would not be able fully to understand the philosophy of it, and that is the vaporous condition of the fuel. In order that you may understand that, let me show you a very pretty but very commonplace experiment. If you blow a can-

dle out cleverly, you will see the vapor rise from it. You have, I know, often smelt the vapor of a blown-out candle, and a very bad smell it is; but if you blow it out cleverly, you will be able to see pretty well the vapor into which this solid matter is transformed. I will blow out one of these candles in such a way as not to disturb the air around it by the continuing action of my breath; and now, if I hold a lighted taper two or three inches from the wick, you will observe a train of fire going through the air till it reaches the candle. I am obliged to be quick and ready, because if I allow the vapor time to cool, it becomes condensed into a liquid or solid, or the stream of combustible matter gets disturbed.

Now as to the shape or form of the flame. It concerns us much to know about the condition which the matter of the candle finally assumes at the top of the wick, where you have such beauty and brightness as nothing but combustion or flame can produce. You have the glittering beauty of gold and silver, and the still higher lustre of jewels like the ruby and diamond; but none of these rival the brilliancy and beauty of flame. What diamond can shine like flame? It owes its lustre at night-time to the very flame shining upon it. The flame shines in darkness, but the light which the diamond has is as nothing until the flame shines upon it, when it is brilliant again. The candle alone shines by itself and for itself, or for those who have arranged the materials. Now let us look a little at the form of the flame as you see it under the glass shade. It is steady and equal, and its general form is that which is reprepainted in the diagram, varying with atmospheric disturbances, and also varying according to the size of the candle. It is a bright oblong, brighter at the top than toward the bottom, with the wick in the middle, and, besides the wick in the middle, certain darker parts toward the bottom, where the ignition is not so perfect as in the part above. I have a drawing here, sketched many years ago by Hooker, when he made his investigations. It is the drawing of the flame of a lamp, but it will apply to the flame of a candle. The cup of the candle is the vessel or lamp; the melted spermaceti is the oil; and the wick is common to both. Upon that he sets this little flame, and then he represents what is true, a certain quantity of matter rising about it which you do not see, and which, if you have not been here before, or are not familiar with the subject, you will not know of. He has here represented the parts of the surrounding atmosphere that are very essential to the flame, and that are always present with it. There is a current formed, which draws the flame out; for the flame which you see is really drawn out by the current, and drawn upward to a great height, just as Hooker has here shown you by that prolongation of the current in the diagram. You may see this by taking a lighted candle, and putting it in the sun so as to get its shadow thrown on a piece of paper. How remarkable it is that that thing which is light enough to produce shadows of other objects can be made to throw its own shadow on a piece of white paper or card, so that you can actually see streaming round the flame something which is not part of the flame, but is ascending and drawing the flame upward. Now I am going to imitate the sunlight by applying the voltaic battery to the electric lamp. You now see our sun and its great luminosity; and by placing a candle between it and the screen, we get the shadow of the flame. You observe the shadow of the candle and of the wick; then there is a darkish part, as represented in the diagram, and then a part which is more distinct. Curiously enough, however, what we see in the shadow as the darkest part of the flame is, in reality, the brightest part; and here you see streaming upward the ascending current of hot air, as shown by Hooker, which draws out the flame, supplies it with air, and cools the sides of the cup of melted fuel.

I can give you here a little farther illustration, for the purpose of showing you how flame goes up or down according to the current. I have here a flame—it is not a candle flame—but you can, no doubt, by this time generalize enough to be able to compare one thing with another: what I am about to do is to change the ascending current that takes the flame upward into a descending current. This I can easily do by the little apparatus you see before me. The flame, as I have said, is not a candle flame, but it is produced by alcohol, so that it shall not smoke too much. I will also color the flame with another substance, so that you may trace its course; for, with the spirit alone, you could hardly see well enough to have the opportunity of tracing its direction. By lighting this spirit of wine we have then a flame produced, and you observe that when held in the air it naturally goes upward. You understand now, easily enough, why flames go up under ordinary circumstances: it is because of the draught of air by which the combustion is formed. But now, by blowing the flame down, you see I am enabled to make it go downward into this little chimney, the direction of the current being changed. . . . You see, then, that we have the power in this way of varying the flame in different directions.

There are now some other points that I must bring before you. Many of the flames you see here vary very much in their shape by the currents of air blowing around them in different directions; but we can, if we like, make flames so that they will look like fixtures, and we can photograph them—indeed, we have to photograph them—so that they become fixed to us, if we wish to find out every thing concerning them. That, however, is not the only thing I wish to mention. If I take a flame sufficiently large, it does not keep that homogeneous, that uniform condition of shape, but it breaks out with a power of life which is quite wonderful. I am about to use another kind of fuel, but one which is truly and fairly a representative of the wax or tallow of a candle. I have here a large ball of cotton, which will serve as a wick. And, now that I have immersed it in spirit and applied a light to it, in what way does it differ from an ordinary candle? Why, it differs very much is one respect, that we have a vivacity and power about it, a beauty and a life entirely different from the light presented by a candle. You see those fine tongues of flame rising up. You have the same general disposition of the mass of the flame from below upward, but, in addition to that, you have this remarkable breaking out into tongues which you do not perceive in the case of a candle. Now, why is this? I must explain it to you, because, when you understand that perfectly, you will be able to follow me better in what I have to say hereafter. I suppose some here will have made for themselves the experiment I am going to show you. Am I right in supposing that any body here has played at snapdragon? I do not know a more beautiful illustration of the philosophy of flame, as to a certain part of its history, than the game of snapdragon. First, here is the dish; and let me say, that when you play snap-dragon properly you ought to have the dish well warmed; you ought also to have warm plums, and warm brandy, which, how-ever, I have not got. When you have put the spirit into the dish, you have the cup and the fuel; and are not the raisins acting like the wicks? I now throw the plums into the dish, and light the spirit, and you see those beautiful tongues of flame that I refer to. You have the air creeping in over the edge of the dish forming these tongues. Why? Because, through the force of the current and the irregularity of the action of the flame, it can not flow in one uniform stream. The air flows in so irregularly that you have what would otherwise be a single image broken up into a variety of forms, and each of these little tongues has an independent exist-ence of its own. Indeed, I might say, you have here a multitude of independent candles. You must not imagine, because you see these tongues all at once, that the flame is of this particular shape. A flame of that shape is never so at any one time. Never is a body of flame, like that which you just saw rising from the ball, of the shape it appears to you. It consists of a multitude of different shapes, succeeding each other so fast that the eye is only able to take cognizance of them all at once. In former times I purposely analyzed a flame of that general character, and the diagram shows you the different parts of which it is composed. They do not occur all at once; it is only because we see these shapes in such rapid succession that they seem to us to exist all at one time.

It is too bad that we have not got farther than my game of snap-dragon; but we must not, under any circumstances, keep you be-yond your time. It will be a lesson to me in future to hold you more strictly to the philosophy of the thing than to take up your time so much with these illustrations.

QUESTIONS AND SUGGESTIONS

1. Faraday takes pains to debunk the idea that science is a mysterious activity. He starts with utterly mundane objects and proceeds to show what the scientist can see that you or I might not see. Imagine yourself attending a Friday Evening Discourse. What general ideas would you carry away about the nature of science?

2. Faraday addresses his audience directly. Can you make use of his model in written discourse? Of his method of referring to visual examples?

3. Faraday claims the privilege of talking to juveniles as a juvenile. Does he patronize his audience? Is he completely straightforward when he says there is no better way of entering science than through studying a candle? How does he organize his material in the light of this self-assumed juvenile *persona*?

4. Faraday uses vivid examples: illustrating capillary attraction by sending a blue saturated salt solution up a column of hardened salt; explaining why a towel dries your hands. How does this use of examples compare with George Gamow's? With J. B. S. Haldane's?

5. Describe and illustrate the properties of various members of a class of objects, imitating Faraday's model exactly or adopting it in whatever way best suits you. Your subject may be manufactured objects like candles or natural ones. In the former case, describing them would include reference to the process of their manufacture, and, in the latter, perhaps reference to their formation, evolution, or genesis.

J. B. S. HALDANE

On Being the Right Size

J. B. S. HALDANE (1892–1964), an Englishman, is famous for estimating the rate of mutation of a human gene and more broadly for the application of mathematical analysis to genetic phenomena. In other aspects of his work on human physiology—for example, on sunstroke and on the "bends"—he often acted as his own experimental animal. He became a Communist in the 1930s, was active in the aid of Nazi refugees, and went into self-exile in India in 1957 out of dissatisfaction with British policies. He was elected a Fellow of the Royal Society in 1932, and, "in recognition of his initiation of the modern phase of study of the evolution of living populations," he received its Darwin medal in 1953. This essay on scale in biology is often cited as an example of the way in which science can change how we look at things. It and the other essays in the volume from which it comes, *Possible Worlds* (1928), were written "in the intervals of research work and teaching, to a large extent in railway trains"; most were first published in periodicals in England and America. Haldane wrote them out of the conviction that "the public has a right to know what is going on inside the laboratories, for some of which it pays."

The most obvious differences between different animals are differences of size, but for some reason the zoologists have paid singularly little attention to them. In a large textbook of zoology before me I find no indication that the eagle is larger than the sparrow, or the hippopotamus bigger than the hare, though some grudging admissions are made in the case of the mouse and the whale. But yet it is easy to show that a hare could not be as large as a hippopotamus, or a whale as small as a herring. For every type of animal there is a most convenient size, and a large change in size inevitably carries with it a change of form.

Let us take the most obvious of possible cases, and consider a giant man sixty feet high—about the height of Giant Pope and Giant Pagan in the illustrated *Pilgrim's Progress* of my childhood. These monsters were not only ten times as high as Christian, but ten times as wide and ten times as thick, so that their total weight was a thousand times his, or about eighty to ninety tons. Unfortunately the cross sections of their bones were only a hundred times

21

those of Christian, so that every square inch of giant bone had to support ten times the weight borne by a square inch of human bone. As the human thigh-bone breaks under about ten times the human weight, Pope and Pagan would have broken their thighs every time they took a step. This was doubtless why they were sitting down in the picture I remember. But it lessens one's respect for Christian and Jack the Giant Killer.

To turn to zoology, suppose that a gazelle, a graceful little creature with long thin legs, is to become large, it will break its bones unless it does one of two things. It may make its legs short and thick, like the rhinoceros, so that every pound of weight has still about the same area of bone to support it. Or it can compress its body and stretch out its legs obliquely to gain stability, like the giraffe. I mention these two beasts because they happen to belong to the same order as the gazelle, and both are quite successful mechanically, being remarkably fast runners.

Gravity, a mere nuisance to Christian, was a terror to Pope, Pagan, and Despair. To the mouse and any smaller animal it presents practically no dangers. You can drop a mouse down a thousand-yard mine shaft; and, on arriving at the bottom, it gets a slight shock and walks away, provided that the ground is fairly soft. A rat is killed, a man is broken, a horse splashes. For the resistance presented to movement by the air is proportional to the surface of the moving object. Divide an animal's length, breadth, and height each by ten; its weight is reduced to a thousandth, but its surface only to a hundredth. So the resistance to falling in the case of the small animal is relatively ten times greater than the driving force.

An insect, therefore, is not afraid of gravity; it can fall without danger, and can cling to the ceiling with remarkably little trouble. It can go in for elegant and fantastic forms of support like that of the daddy-longlegs. But there is a force which is as formidable to an insect as gravitation to a mammal. This is surface tension. A man coming out of a bath carries with him a film of water of about one-fiftieth of an inch in thickness. This weighs roughly a pound. A wet mouse has to carry about its own weight of water. A wet fly has to lift many times its own weight and, as everyone knows, a fly once wetted by water or any other liquid is in a very serious position indeed. An insect going for a drink is in as great danger as a man leaning out over a precipice in search of food. If it once falls into the grip of the surface tension of the water—that is to say, gets wet—it is likely to remain so until it drowns. A few insects, such as water-beetles, contrive to be unwettable; the majority keep well away from their drink by means of a long proboscis.

Of course tall land animals have other difficulties. They have to pump their blood to greater heights than a man, and, therefore, require a larger blood pressure and tougher blood-vessels. A great many men die from burst arteries, especially in the brain, and this danger is presumably still greater for an elephant or a giraffe. But animals of all kinds find difficulties in size for the following reason. A typical small animal, say a microscopic worm or rotifer, has a smooth skin through which all the oxygen it requires can soak in, a straight gut with sufficient surface to absorb its food, and a single kidney. Increase its dimensions tenfold in every direction, and its weight is increased a thousand times, so that if it is to use its muscles as efficiently as its miniature counterpart, it will need a thousand times as much food and oxygen per day and will excrete a thousand times as much of waste products.

Now if its shape is unaltered its surface will be increased only a hundredfold, and ten times as much oxygen must enter per minute through each square millimetre of skin, ten times as much food through each square millimetre of intestine. When a limit is reached to their absorptive powers their surface has to be increased by some special device. For example, a part of the skin may be drawn out into tufts to make gills or pushed in to make lungs, thus increasing the oxygen-absorbing surface in proportion to the animal's bulk. A man, for example, has a hundred square yards of lung. Similarly, the gut, instead of being smooth and straight, becomes coiled and develops a velvety surface, and other organs increase in complication. The higher animals are not larger than the lower because they are more complicated. They are more complicated because they are larger. Just the same is true of plants. The simplest plants, such as the green algae growing in stagnant water or on the bark of trees, are mere round cells. The higher plants increase their surface by putting out leaves and roots. Comparative anatomy is largely the story of the struggle to increase surface in proportion to volume.

Some of the methods of increasing the surface are useful up to

FRANCIS CRICK

The Uniformity of Biochemistry

FRANCIS CRICK (b. 1916) is a professor at the Salk Institute for Biological Studies in San Diego. In 1962 he won the Nobel Prize for physiology or medicine with James Watson for the discovery of the double helix. Crick is the author of numerous articles and two recent books, *What Mad Pursuit: A Personal View of Scientific Discovery* (1988) and *Life Itself: Its Origin and Nature* (1981).

The problem of the origin of life is, at bottom, a problem in organic chemistry—the chemistry of carbon compounds—but organic chemistry within an unusual framework. Living things, as we shall see, are specified in detail at the level of atoms and molecules, with incredible delicacy and precision. At the beginning it must have been molecules that evolved to form the first living system. Because life started on earth such a long time ago—perhaps as much as four billion years ago—it is very difficult for us to discover what the first living things were like. All living things on earth, without exception, are based on organic chemistry, and such chemicals are usually not stable over very long periods of time at the range of temperatures which exist on the earth's surface. The constant buffeting of thermal motion over hundreds of millions of years eventually disrupts the strong chemical bonds which hold the atoms of an organic molecule firmly together over shorter periods; over our own lifetime, for example. For this reason it is almost impossible to find "molecular fossils" from these very early times.

Minerals can be much more stable, at least on a somewhat coarser scale, mainly because their atoms use strong bonds to form regular three-dimensional structures. The failure of a single bond will not disturb the shape of the mineral too much. Fossils are seen in abundance in rocks laid down a little over half a billion years ago, at a time when organisms had evolved sufficiently to develop hard parts. Such fossils are not usually made of the original material of those organisms but consist of mineral deposits which have infiltrated them and taken up their shape. The shape of the soft parts is usually lost, though occasionally traces like wormholes are preserved—footprints on the rocks of time.

Are there any fossils much earlier than this? Careful microscopic examination of very early rocks has shown them to contain small structures which look like the fossilized remnants of very simple organisms, rather similar to some of the unicellular organisms on the earth today. This makes good sense. In the process of evolution we would expect creatures with many cells to develop from earlier ones having only single cells. Although there is still some controversy about the details, the earliest organisms of this type have been dated to about $2\frac{1}{2}$ to $3\frac{1}{2}$ billion years ago. The age of the earth is about $4\frac{1}{2}$ billion years. After the turmoil of its initial formation had subsided there was a period of about a billion years during which life could have evolved from the complex chemistry of the earth's surface, especially in its oceans, lakes, and pools. Of that period we have no fossil record at all, because no preserved parts of the sedimentary rocks from that time have yet been found.

There are only two ways for us to approach this problem. We can try to simulate those early conditions in the laboratory. Since life is probably a happy accident which, even in the extended laboratory of the planet's surface, is likely to have taken many millions of years to occur, it is not too surprising that such research has not yet got very far, though some progress has been made. In addition we can look carefully at all living things which exist today. Because they are all descended from some of the first simple organisms, it might be hoped that they still bear within them some traces of the earliest living things.

At first sight such a hope seems absurd. What could possibly unite the lily and the giraffe? What could a man share with the bacteria in his intestines? A cynic might wonder whether, since all living things eat or are eaten, this at least suggests they have something in common. Remarkably, this turns out to be correct. The unity of biochemistry is far greater and more detailed than was supposed even as little as a hundred years ago. The immense variety of nature—man, animals, plants, microorganisms, even vi-

ruses—is built, at the chemical level, on a common ground plan.
It is the fantastic elaboration of this ground plan, evolved by natural selection over countless generations, which makes it difficult
for us, in our everyday life, to penetrate beneath the outward form
and perceive the unity within. In spite of our differences we all
use a single chemical language, or, more precisely, as we shall see,
two such languages, intimately related to each other.

To understand the unity of biochemistry we must first grasp
in a very general way what chemical reactions go on within an
organism. A living cell can be thought of as a fairly complex, well-
organized chemical factory which takes one set of organic molecules—its food—breaks them down, if necessary, into smaller units
and then reassorts and recombines these smaller units, often in
several discrete steps, to make many other small molecules, some
of which it excretes and some of which it uses for further synthesis.
In particular, it strings special sets of these small molecules together into long chains, usually unbranched, to make the vital
macromolecules of the cell, the three great families of giant molecules: the nucleic acids, the proteins, and the polysaccharides.

The first level of organization we must consider is the lowest
of all—that at which atoms are bound together to form small molecules. Now, a single atom is a fairly symmetrical object. Its shape
is approximately spherical and if we look at it in a mirror it appears exactly the same, just as a billiard ball would. More intricate
structures can have a "handedness"—our own hands are a good
example. If we look at a right hand in a mirror we see a left hand,
and vice versa. We can oppose our two hands, as in prayer, but
this is as if we held a mirror between them. There is no way in
which we can exactly superimpose one on the other, even in our
imagination.

Some simple organic molecules, such as alcohol, have no "hand";
they are identical to their mirror images, as indeed a cup is. But
this is not true of most organic molecules. The sugar on the breakfast table, if looked at in a mirror, becomes a significantly different
assembly of atoms. This difference does not matter for *all* types
of chemical reaction. If we heated such a molecule and could watch
the molecular vibrations increase until one of the bonds broke,
we would see that, had we imagined the mirror image of this
process, the relative movements of all the atoms would have been

identical. The basic reactions of chemistry are symmetrical under reflection to a very high degree of approximation. The difference in the hand only becomes important when two molecules have to fit together. We can see this in the manufacture of a glove. All the components of a glove—the fabric, the sewing thread, even the buttons—are, individually, mirror-symmetric, but they can be put together in two similar but different ways, to make either a right-handed glove or a left-handed one. Obviously we need two sorts because we have two kinds of hands—a good left-handed glove will not fit properly onto a right hand.

The simplest form of asymmetrical molecule of this type arises when a single carbon atom is joined by single bonds to four other *different* atoms, or groups of atoms. This is because the four bonds of the carbon atom do not all lie in the same plane but are spaced out equally in all three dimensions, pointing approximately toward the corners of a regular tetrahedron.

Thus, organic molecules—molecules containing carbon atoms—often have a hand, even though they may be small, but we still have to realize why this matters in a cell. The basic reason is that a biochemical molecule does not exist in isolation. It reacts with other molecules. Almost every biochemical reaction is speeded up by its own special catalyst. A small molecule, to react in this way, has to fit snugly onto the catalyst's surface, and since the small molecule has a hand, the catalyst must also have one. As in the case of a glove, the reaction will not work properly if we try to

The distribution in space of the four bonds around a single carbon atom.

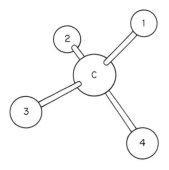

made to deduce the relationship between the two languages from chemical principles, but so far none have been successful. The code has a few regular features, but these might be due to chance.

Even if there existed an entirely separate form of life elsewhere, also based on nucleic acids and protein, I can see no good reason why the genetic code should be exactly the same there as it is here. (The Morse Code, incidentally, is not completely arbitrary. The commonest letters, like *e* and *t*, are allocated the shortest number of dots or dashes.) If this appearance of arbitrariness in the genetic code is sustained, we can only conclude, once again, that all life on earth arose from one very primitive population which first used it to control the flow of chemical information from the nucleic acid language to the protein language.

Thus, all living things use the same four-letter language to carry genetic information. All use the same twenty-letter language to construct their proteins, the machine tools of the living cell. All use the same chemical dictionary to translate from one language to the other. Such an astonishing degree of uniformity was hardly suspected as little as forty years ago, when I was an undergraduate. I find it a curious symptom of our times that those who derive deep satisfaction from brooding on their unity with nature are often quite ignorant of the very unity they are attempting to contemplate. Perhaps in California there already exists a church in which the genetic code is read out every Sunday morning, though I doubt whether anyone would find such a bare recital very inspiring.

We see, then, that one way to approach the origin of life is to try to imagine how this remarkable uniformity first arose. Almost all modern theories and experimental work on life's origin take as their starting point the synthesis of either nucleic acid or protein or both. How could the primitive earth (if indeed life first started on earth) have produced the first relevant macromolecules? We have seen that these chain molecules are made by joining together small subunits end to end. How could the small molecules have been synthesized under early, prebiotic conditions? And how could we decide, even if we could have watched the whole operation in atomic detail, when the system first deserved to be called "living"? To come to grips with this problem we must examine just what attributes we would expect *any* living system to have.

QUESTIONS AND SUGGESTIONS

1. For what type of audience is Crick writing? Give examples to support your estimation. How do the style and content compare to Bateson's essay?

2. How is uniformity achieved in biochemistry?

3. After reading the individual selections by Watson and Crick and the jointly authored essay, what conclusions can you draw about these scientists' approaches to writing about science?

4. The selections by Crick, Watson, Bateson, Margulis, Haldane, and Huxley deal with various aspects of genetics and biochemistry. Organize these essays into a progressive unit and address the evolution of the history of genetic thought as manifested by these authors in an essay of 750 words or longer.

ISAAC ASIMOV

Organic Synthesis

ISAAC ASIMOV (b. 1920) says that he has "long since decided to make it a rule to cease listing my books," but their number is definitely large. He has written popular surveys of many fields of medicine, science, and mathematics. He writes science fiction and is the unofficial wizard of the sci-fi world. Under the pseudonym of Paul French, he writes *more* science fiction. He came to America at the age of three and later went to Columbia University. After some years on the faculty of Boston University's School of Medicine he left to devote his full time to writing, his consuming passion. This selection, from *An Intelligent Man's Guide to Science*, illustrates his functional style of science writing. He wants to convey central concepts in a historical context, and he mixes in biographical anecdotes and industrial history for good measure, to keep the "intelligent man" interested.

After Wöhler had produced urea from ammonium cyanate and chemists had formed various other organic molecules by trial and error, in the 1850's there came a chemist who went systematically and methodically about the business of synthesizing organic subtances in the laboratory. He was the Frenchman Pierre Eugène Marcelin Berthelot. He prepared a number of simple organic compounds from still simpler inorganic compounds such as carbon monoxide. Berthelot built his simple organic compounds up through increasing complexity until he finally had ethyl alcohol, among other things. It was "synthetic ethyl alcohol," to be sure, but absolutely indistinguishable from the "real thing," because it *was* the real thing.

Ethyl alcohol is an organic compound familiar to all and highly valued by most. No doubt the thought that the chemist could make ethyl alcohol from coal, air, and water (coal to supply the carbon, air the oxygen, and water the hydrogen), without the necessity of fruits or grain as a starting point, must have created enticing visions and endowed the chemist with a new kind of reputation as a miracle worker. At any rate, it put organic synthesis on the map.

For chemists, however, Berthelot did something even more significant. He began to form products that did not exist in nature. He took "glycerol," a component obtained from the breakdown of the fats of living organisms, and combined it with acids not known to occur naturally in fats (although they occurred naturally elsewhere). In this way he obtained fatty substances which were not quite like those that occurred in organisms.

Thus Berthelot laid the groundwork for a new kind of organic chemistry—the synthesis of molecules that nature could not supply. This meant the possible formation of a kind of "synthetic" which might be a substitute—perhaps an inferior substitute—for some natural compound that was hard or impossible to get in the needed quantity. But it also meant the possibility of "synthetics" which were improvements on anything in nature.

This notion of improving on nature in one fashion or another, rather than merely supplementing it, has grown to colossal proportions since Berthelot showed the way. The first fruits of the new outlook were in the field of dyes.

The beginnings of organic chemistry were in Germany. Wöhler and Liebig were both German, and other men of great ability followed them. Before the middle of the nineteenth century, there were no organic chemists in England even remotely comparable to those in Germany. In fact, English schools had so low an opinion of chemistry that they taught the subject only during the lunch recess, not expecting (or even perhaps desiring) many students to be interested. It is odd, therefore, that the first feat of synthesis with world-wide repercussions was actually carried through in England.

It came about in this way. In 1845, when the Royal College of Science in London finally decided to give a good course in chemistry, it imported a young German to do the teaching. He was August Wilhelm von Hofmann, only 27 at the time, and he was hired at the suggestion of Queen Victoria's husband, the Prince Consort Albert (who was himself of German birth).

Hofmann was interested in a number of things, among them coal tar, which he had worked with on the occasion of his first research project under Liebig. Coal tar is a black, gummy material given off by coal when it is heated strongly in the absence of air.

The tar is not an attractive material, but it is a valuable source of organic chemicals. In the 1840's, for instance, it served as a source of large quantities of reasonably pure benzene and of a nitrogen-containing compound called "aniline" which is related to benzene.

About ten years after he arrived in England, Hofmann came across a 17-year-old boy studying chemistry at the college. His name was William Henry Perkin. Hofmann had a keen eye for talent and knew enthusiasm when he saw it. He took on the youngster as an assistant and set him to work on coal-tar compounds. Perkin's enthusiasm was tireless. He set up a laboratory in his home and worked there as well as at school.

Hofmann, who was also interested in medical applications of chemistry, mused aloud one day in 1856 on the possibility of synthesizing quinine, a natural substance used in the treatment of malaria. Now those were the days before structural formulas had come into their own. The only thing known about quinine was its composition, and no one at the time had any idea of just how complicated its structure was.

Blissfully ignorant of its complexity, Perkin, at the age of 18, tackled the problem of synthesizing quinine. He began with allyltoluidine, one of his coal-tar compounds. This molecule seemed to have about half the numbers of the various types of atoms that quinine had in its molecule. If he put two of these molecules together and added some missing oxygen atoms (say by mixing in some potassium dichromate, known to add oxygen atoms to chemicals with which it was mixed), Perkin thought he might get a molecule of quinine.

Naturally this approach got Perkin nowhere. He ended with a dirty, red-brown goo. Then he tried aniline in place of allyltoluidine and got a blackish goo. This time, though, it seemed to him that he caught a purplish glint in it. He added alcohol to the mess, and the colorless liquid turned a beautiful purple. At once Perkin thought of the possibility that he had discovered something that might be useful as a dye.

Dyes had always been greatly admired, and expensive, substances. There were only a handful of good dyes—dyes that stained fabric permanently and brilliantly and did not fade or wash out. There was dark blue indigo, from the indigo plant; there was "Tyrian purple," from a snail (so-called because ancient Tyre grew rich on its manufacture—in the later Roman Empire the royal children were born in a room with hangings dyed with Tyrian purple, whence the phrase "born to the purple"); and there was reddish alizarin, from the madder plant ("alizarin" came from Arabic words meaning "the juice"). To these inheritances from ancient and medieval times later dyers had added a few tropical dyes and inorganic pigments (today used chiefly in paints).

This explains Perkin's excitement about the possibility that his purple substance might be a dye. At the suggestion of a friend, he sent a sample to a firm in Scotland which was interested in dyes, and quickly the answer came back that the purple compound had good properties. Could it be supplied cheaply? Perkin proceeded to patent the dye (there was considerable argument as to whether an 18-year-old could obtain a patent, but eventually he obtained it), to quit school, and to go into business.

His project wasn't easy. Perkin had to start from scratch, preparing his own starting materials from coal tar with equipment of his own design. Within six months, however, he was producing what he named "Aniline Purple"—a compound not found in nature and superior to any natural dye in its color range.

French dyers, who took to the new dye more quickly than did the more conservative English, named the color "mauve," from the mallow (Latin name "malva"), and the dye itself came to be known as "mauveine." Quickly it became the rage (the period being sometimes referred to as the "Mauve Decade"), and Perkin grew rich. At the age of 23 he was the world authority on dyes.

The dam had broken. A number of organic chemists, inspired by Perkin's astonishing success, went to work synthesizing dyes, and many succeeded. Hofmann himself turned to this new field, and in 1858 he synthesized a red-purple dye which was later given the name "magenta" by the French dyers (then, as now, arbiters of the world's fashions). The dye was named for the Italian city where the French defeated the Austrians in a battle in 1859.

Hofmann returned to Germany in 1865, carrying his new interest in dyes with him. He discovered a group of violet dyes still known as "Hofmann's violets."

Chemists also synthesized the natural dyestuffs in the laboratory. Karl Graebe of Germany and Perkin both synthesized alizarin in 1869 (Graebe applying for the patent one day sooner than Perkin), and in 1880 the German chemist Adolf von Baeyer worked out a method of synthesizing indigo. (For his work on dyes von Baeyer received the Nobel Prize in chemistry in 1905.)

Perkin retired from business in 1874, at the age of 35, and returned to his first love, research. By 1875 he had managed to synthesize coumarin (a naturally-occurring substance which has the pleasant odor of new-mown hay); this served as the beginning of the synthetic perfume industry.

Perkin alone could not maintain British supremacy against the great development of German organic chemistry, and by the turn of the century "synthetics" had become virtually a German monopoly. But during World War I, Great Britain and the United States, shut off from the products of the German chemical laboratories, were forced to develop chemical industries of their own.

Achievements in synthetic organic chemistry could not have proceeded at anything better than a stumbling pace if chemists had had to depend upon fortunate accidents such as the one that had been seized upon by Perkin. Fortunately the structural formulas of Kekulé, presented three years after Perkin's discovery, made it possible to prepare blueprints, so to speak, of the organic molecule. No longer did chemists have to try to prepare quinine by sheer guesswork and hope; they had methods for attempting to scale the structural heights of the molecule step by step, with advance knowledge of where they were headed and what they might expect.

Chemists learned how to alter one group of atoms to another; to open up rings of atoms and to form rings from open chains; to split groups of atoms in two, and to add carbon atoms one by one to a chain. The specific method of doing a particular architectural task within the organic molecule is still often referred to by the name of the chemist who first described the details. For instance, Perkin discovered a method of adding a two-carbon atom group by heating certain substances with chemicals named acetic anhydride and sodium acetate. This is still called the "Perkin Reaction." Perkin's teacher, Hofmann, discovered that a ring of atoms which included a nitrogen could be treated with a substance called methyl iodide in the presence of silver compound in such a way that the ring was eventually broken and the nitrogen atom removed. This is the "Hofmann Degradation." In 1877 the French chemist Charles Friedel, working with the American chemist James Mason Crafts, discovered a way of attaching a short carbon chain to a benzene ring by the use of heat and aluminum chloride. This is now known as the "Friedel-Crafts Reaction."

In 1900 the French chemist Victor Grignard discovered that magnesium metal, properly used, could bring about a rather large variety of different joinings of carbon chains. For the development of these "Grignard Reactions" he shared in the Nobel Prize in chemistry in 1912. The French chemist Paul Sabatier, who shared it with him, had discovered (with J. B. Senderens) a method of using finely divided nickel to bring about the addition of hydrogen atoms in those places where a carbon chain possessed a double bond. This is the "Sabatier-Senderens Reduction."

In other words, by noting the changes in the structural formulas of substances subjected to a variety of chemicals and conditions, organic chemists worked out a slowly growing set of ground rules on how to change one compound into another at will. It wasn't easy. Every compound and every change had its own peculiarities and difficulties. But the main paths were blazed, and the skilled organic chemist found them clear signs toward progress in what had formerly seemed a jungle.

Knowledge of the manner in which particular groups of atoms behaved could also be used to work out the structure of unknown compounds. For instance, when simple alcohols react with metallic sodium and liberate hydrogen, only the hydrogen linked to an oxygen atom is released, not the hydrogens linked to carbon atoms. On the other hand, some organic compounds will take on hydrogen atoms under appropriate conditions while others will not. It turns out that compounds that add hydrogen generally possess double or triple bonds and add the hydrogen at those bonds. From such information a whole new type of chemical analysis of organic compounds arose; the nature of the atom-groupings was determined, rather than just the numbers and kinds of various atoms present. The liberation of hydrogen by the addition of sodium signified the presence of an oxygen-bound hydrogen atom in the compound; the acceptance of hydrogen meant the presence of double or triple bonds. If the molecule was too complicated for analysis as a whole, it could be broken down into simpler portions by well-defined methods; the structures of the simpler portions could be worked out and the original molecule deduced from those.

Using the structural formula as a tool and guide, chemists could work out the structure of some useful naturally occurring organic compound (analysis) and then set about duplicating it or something like it in the laboratory (synthesis). One result was that something which was rare, expensive or difficult to obtain in nature might become cheaply available in quantity in the laboratory. Or, as in the case of the coal-tar dyes, the laboratory might create something that fulfilled a need better than did similar substances found in nature.

One startling case of a deliberate improvement on nature involves the drug cocaine. Cocaine is found in the leaves of the coca plant, which is native to Bolivia and Peru (but is now grown chiefly in Java). The South American Indians would chew coca leaves, finding it an antidote to fatigue and a source of happiness-sensation. The Scottish physician Sir Robert Christison introduced the plant to Europe, and eventually cocaine was isolated as the active principle. In 1884 the American physician Carl Koller discovered that cocaine could be used as a local anesthetic when added to the mucous membranes around the eye. Eye operations could then be performed without pain. Cocaine could also be used in dentistry, allowing teeth to be extracted without pain.

Anesthetics had come into general use about 40 years before that. The American surgeon Crawford Williamson Long in 1842 had used ether to put a patient to sleep during tooth extractions. In 1846 the American dentist William Thomas Green Morton conducted a surgical operation under ether at the Massachusetts General Hospital. Morton usually gets the credit for the discovery, because Long did not describe his feat in the medical journals until after Morton's public demonstration. In any case, doctors were quite aware that anesthesia had finally converted surgery from torture-chamber butchery to something that was at least humane and, with the addition of antiseptic conditions, even life-saving. For that reason any further advance in anesthesia was seized upon with great interest, and this included cocaine.

There were several drawbacks to cocaine. In the first place, it induced troublesome side-effects and could even kill patients sensitive to it. Secondly, it could bring about addiction and had to be used skimpily and with caution. (Cocaine is one of the dangerous "dopes." Up to 20 tons of it are produced illegally each year and sold with tremendous profits to a few and tremendous misery to many, despite world-wide efforts to stop the traffic.) Thirdly, the molecule is fragile, and heating cocaine to sterilize it of any bacteria leads to changes in the molecule that interfere with its anesthetic effects.

The structure of the cocaine molecule is rather complicated:

The double ring on the left is the fragile portion, and that is the difficult one to synthesize. (The synthesis of cocaine wasn't achieved until 1923, when the German chemist Richard Willstätter managed it.) However, it occurred to chemists that they might synthesize similar compounds in which the double ring was not closed. This would make the compound both easier to form and more stable. The synthetic substance might possess the anesthetic properties of cocaine, perhaps without the undesirable side-effects.

For some 20 years German chemists tackled the problem, turning out dozens of compounds, some of which were pretty good. The most successful modification was obtained in 1909, when a compound with the following formula was prepared:

Compare this with the formula for cocaine and you will see the similarity, and also the important fact that the double ring no longer exists. This simpler molecule—stable, easy to synthesize, with good anesthetic properties and very little in the way of side-effects—does not exist in nature. It is a "synthetic substitute" far better than the real thing. It is called "procaine," but is better known to the public by the trade-name Novocaine.

A series of other anesthetics have been synthesized in the half-century since, and, thanks to chemistry, doctors and dentists have an assortment of effective and safe pain-killers at hand.

Man now has at his disposal all sorts of synthetics of great potential use and misuse: explosives, poison gases, insecticides, weed-killers, antiseptics, disinfectants, detergents, drugs—almost no end of them, really. But synthesis is not merely the handmaiden of consumer needs. It can also be placed at the service of pure chemical research.

It often happens that a complex compound, produced either by living tissue or by the apparatus of the organic chemist, can only be assigned a tentative molecular structure, after all possible deductions have been drawn from the nature of the reactions it undergoes. In that case, a way out is to synthesize a compound by means of reactions designed to yield a molecular structure like the one that has been deduced. If the properties of the resulting compound are identical with the compound being investigated in the first place, the assigned structure becomes something in which a chemist can place his confidence.

An impressive case in point involves hemoglobin, the main component of the red blood cells and the pigment that gives the blood its red color. In 1831 the French chemist L. R. LeCanu split hemoglobin into two parts, of which the smaller portion, called "heme," made up 4 per cent of the mass of hemoglobin. Heme was found to have the empirical formula $C_{34}H_{32}O_4N_4Fe$. Compounds like heme were known to occur in other vitally important substances, both in the plant and animal kingdoms, and so the structure of the molecule was a matter of great moment to biochemists. For nearly a century after LeCanu's isolation of heme, however, all that could be done was to break it down into smaller molecules. The iron atom (Fe) was easily removed, and what was left then broke up into pieces roughly a quarter the size of the original molecule. These fragments were found to be "pyrroles"—molecules built on rings of five atoms, of which four are carbon and one nitrogen. Pyrrole itself has the following structure:

$$
\begin{array}{ccc}
\text{CH} & \!\!-\!\! & \text{CH} \\
/\!/ & & \backslash\!\backslash \\
\text{CH} & & \text{CH} \\
\backslash & & / \\
& \text{NH} &
\end{array}
$$

The pyrroles actually obtained from heme possessed small groups of atoms containing one or two carbon atoms attached to the ring in place of one or more of the hydrogen atoms.

In the 1920's the German chemist Hans Fischer tackled the problem further. Since the pyrroles were one quarter the size of the original heme, he decided to try to combine four pyrroles and see what he got. What he finally succeeded in getting was a four-ring compound which he called "porphin" (from a Greek word meaning "purple," because of its purple color). Porphin would look like this:

However, the pyrroles obtained from heme in the first place contained small "side-chains" attached to the ring. These remained in place when the pyrroles were joined to form porphin. The porphin with various side-chains attached make up a family of compounds called the "porphyrins." It was obvious to Fischer upon comparing the properties of heme with those of the porphyrins he had synthesized that heme (minus its iron atom) was a porphyrin. But which one? No fewer than 15 different compounds could be formed from the various pyrroles obtained from heme,

ent, any more than the three dimensions of space are. We still need four quantities to determine the position of an event, but we cannot, as before, divide off one of the four as quite independent of the other three.

It is not quite true to say that there is no longer any distinction between time and space. As we have seen, there are time-like intervals and space-like intervals. But the distinction is of a different sort from that which was formerly assumed. There is no longer a universal time which can be applied without ambiguity to any part of the universe; there are only the various "proper" times of the various bodies in the universe, which agree approximately for two bodies which are not in rapid motion, but never agree exactly except for two bodies which are at rest relatively to each other.

The picture of the world which is required for this new state of affairs is as follows: Suppose an event E occurs to me, and simultaneously a flash of light goes out from me in all directions. Anything that happens to any body after the light from the flash has reached it is definitely after the event E in any system of reckoning time. Any event anywhere which I could have seen before the event E occurred to me is definitely before the event E in any system of reckoning time. But any event which happened in the intervening time is not definitely either before or after the event E. To make the matter definite: suppose I could observe a person in Sirius, and he could observe me. Anything which he does, and which I see before the event E occurs to me, is definitely before E; anything he does after he has seen the event E is definitely after E. But anything that he does before he sees the event E, but so that I see it after the event E has happened, is not definitely before or after E. Since light takes many years to travel from Sirius to earth, this gives a period of twice as many years in Sirius which may be called "contemporary" with E, since these years are not definitely before or after E.

Dr. A. A. Robb, in his *Theory of Time and Space,* suggested a point of view which may or may not be philosophically fundamental, but is at any rate a help in understanding the state of affairs we have been describing. He maintained that one event can only be said to be definitely *before* another if it can influence that other in some way. Now influences spread from a centre at varying rates. Newspapers exercise an influence emanating from London at an average rate of about twenty miles an hour—rather more for long distances. Anything a man does because of what he reads in the newspaper is clearly subsequent to the printing of the newspaper. Sounds travel much faster: it would be possible to arrange a series of loud-speakers along the main roads, and have newspapers shouted from each to the next. But telegraphing is quicker, and wireless telegraphy travels with the velocity of light, so that nothing quicker can ever be hoped for. Now what a man does in consequence of receiving a wireless message he does *after* the message was sent; the meaning here is quite independent of conventions as to the measurement of time. But anything that he does while the message is on its way cannot be influenced by the sending of the message, and cannot influence the sender until some little time after he sent the message, that is to say, if two bodies are widely separated, neither can influence the other except after a certain lapse of time; what happens before that time has elapsed cannot affect the distant body. Suppose, for instance, that some notable event happens on the sun: there is a period of sixteen minutes on the earth during which no event on the earth can have influenced or been influenced by the said notable event on the sun. This gives a substantial ground for regarding that period of sixteen minutes on the earth as neither before nor after the event on the sun.

The paradoxes of the special theory of relativity are only paradoxes because we are unaccustomed to the point of view, and in the habit of taking things for granted when we have no right to do so. This is especially true as regards the measurement of lengths. In daily life, our way of measuring lengths is to apply a foot-rule or some other measure. At the moment when the foot-rule is applied, it is at rest relatively to the body which is being measured. Consequently the length that we arrive at by measurement is the "proper" length, that is to say, the length as estimated by an observer who shares the motion of the body. We never, in ordinary life, have to tackle the problem of measuring a body which is in continual motion. And even if we did, the velocities of visible bodies on the earth are so small relatively to the earth that the anomalies dealt with by the theory of relativity would not appear. But in astronomy, or in the investigation of atomic structure, we are faced with problems which cannot be tackled in this

way. Not being Joshua, we cannot make the sun stand still while
we measure it; if we are to estimate its size we must do so while it
is in motion relatively to us. And similarly if you want to estimate
the size of an electron, you have to do so while it is in rapid motion, because it never stands still for a moment. This is the sort of
problem with which the theory of relativity is concerned. Measurement with a foot-rule, when it is possible, gives always the same
result, because it gives the "proper" length of a body. But when
this method is not possible, we find that curious things happen,
particularly if the body to be measured is moving very fast relatively to the observer. A figure like the one [below] will help us
to understand the state of affairs.

Let us suppose that the body on which we wish to measure
lengths is moving relatively to ourselves, and that in one second it
moves the distance OM. Let us draw a circle round O whose radius
is the distance that light travels in a second. Through M draw
MP perpendicular to OM, meeting the circle in P. Thus OP is the
distance that light travels in a second. The ratio of OP to OM is
the ratio of the velocity of light to the velocity of the body. The
ratio of OP to MP is the ratio in which apparent lengths are altered by the motion. That is to say, if the observer judges that two
points in the line of motion on the moving body are at a distance
from each other represented by MP, a person moving with the
body would judge that they were at a distance represented (on the
same scale) by OP. Distances on the moving body at right angles
to the line of motion are not affected by the motion. The whole
thing is reciprocal; that is to say, if an observer moving with the

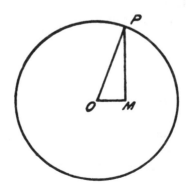

body were to measure lengths on the previous observer's body, they would be altered in just the same proportion. When two bodies are moving relatively to each other, lengths on either appear shorter to the other than to themselves. This is the Fitzgerald contraction, which was first invented to account for the result of the Michelson-Morley experiment. But it now emerges naturally from the fact that the two observers do not make the same judgment of simultaneity.

The way in which simultaneity comes in is this: We say that two points on a body are a foot apart when we can *simultaneously* apply one end of a foot-rule to the one and the other end to the other. If, now, two people disagree about simultaneity, and the body is in motion, they will obviously get different results from their measurements. Thus the trouble about time is at the bottom of the trouble about distance.

The ratio of OP to MP is the essential thing in all these matters. Times and lengths and masses are all altered in this proportion when the body concerned is in motion relatively to the observer. It will be seen that, if OM is very much smaller than OP, that is to say, if the body is moving very much more slowly than light, MP and OP are very nearly equal, so that the alterations produced by the motion are very small. But if OM is nearly as large as OP, that is to say, if the body is moving nearly as fast as light, MP becomes very small compared to OP, and the effects become very great. The apparent increase of mass in swiftly moving particles had been observed, and the right formula had been found, before Einstein invented his special theory of relativity. In fact, Lorentz had arrived at the formulae called the "Lorentz transformation," which embody the whole mathematical essence of the special theory of relativity. But it was Einstein who showed that the whole thing was what we ought to have expected, and not a set of makeshift devices to account for surprising experimental results. Nevertheless, it must not be forgotten that experimental results were the original motive of the whole theory, and have remained the ground for undertaking the tremendous logical reconstruction involved in Einstein's theories.

We may now recapitulate the reasons which have made it necessary to substitute "space-time" for space and time. The old separation of space and time rested upon the belief that there was no

theorems concerning the numerical relationships between various distances and angles (as, for example, the famous Pythagorean theorem concerning the three sides of a right-angled triangle), the fact is that a great many of the most fundamental properties of space do not require any measurements of lengths or angles whatsoever. The branch of geometry concerned with these matters is known as *analysis situs* or *topology*[2] and is one of the most provocative and difficult of the departments of mathematics.

To give a simple example of a typical topological problem, let us consider a closed geometrical surface, say that of a sphere, divided by a network of lines into many separate regions. We can prepare such a figure by locating on the surface of a sphere an arbitrary number of points and connecting them with non-intersecting lines. What are the relationships that exist between the number of original points, the number of lines representing the boundaries between adjacent regions, and the number of regions themselves?

First of all, it is quite clear that if instead of the sphere we had taken a flattened spheroid like a pumpkin, or an elongated body like a cucumber, the number of points, lines, and regions would have been exactly the same on a perfect sphere. In fact, we can take any closed surface that can be obtained by deforming a rubber balloon, by stretching it, by squeezing it, by doing to it anything we like, except cutting or tearing it, and neither the formulation nor the answer to our question will change in the slightest way. This fact presents a striking contrast to the facts of ordinary numerical relationships in geometry (such as the relationships that exist among linear dimensions, surface areas, and volumes of geometrical bodies). Indeed such relationships would be materially distorted if we stretched a cube into a parallel pipe, or squeezed a sphere into a pancake.

One of the things we can do with our sphere divided into a number of separate regions is to flatten each region so that the sphere becomes a polyhedron; the lines bounding different regions now become the edges of the polyhedron, and the original set of points become its vertices.

[2] Which means, from the Latin and the Greek respectively, the study of locations.

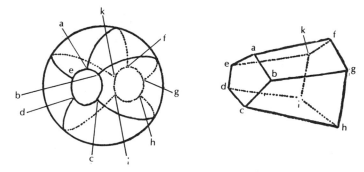

Figure 2.   A subdivided sphere transformed into a polyhedron.

Our previous problem can now be reformulated, without however changing its sense, into a question concerning the relationships between the number of vertices, edges, and faces in a polyhedron of an arbitrary type.

In figure 3 we show five regular polyhedrons, that is, those in
which all faces have an equal number of sides and vertices, and
one irregular one drawn simply from imagination.

In each of these geometrical bodies we can count the number of
vertices, the number of edges, and the number of faces. What is
the relation between these three numbers, if any?

By direct counting we can build the accompanying table.

Name	V number of vertices	E number of edges	F number of faces	V + F	E + 2
Tetrahedron (pyramid)	4	6	4	8	8
Hexahedron (cube)	8	12	6	14	14
Octahedron	6	12	8	14	14
Icosahedron	12	30	20	32	32
Dodecahedron or Pentagon- dodecahedron	20	30	12	32	32
"Monstrosity"	21	45	26	47	47

At first the figures given in the three columns (under V, E, and F) do not seem to show any definite correlation, but after a little study you will find that the sum of the figures in the V and F columns always exceed the figure in the E column by two. Thus we can write the mathematical relationship:

$$V + F = E + 2.$$

Does this relationship hold for only the five particular polyhedrons shown in figure 3, or is it also true for any polyhedron? If you try to draw several other polyhedrons different from those shown in figure 3, and count their vertices, edges, and faces, you will find that the above relationship exists in every case. Apparently then, $V + F = E + 2$ is a general mathematical theorem of a topological nature since the relationship expression does not depend on measuring the lengths of the ribs, or the areas of the faces, but is concerned only with the number of the different geometrical units (that is, vertices, edges, faces) involved.

The relationship we have just found between the number of vertices, edges, and faces in a polyhedron was first noticed by the famous French mathematician of the seventeenth century, René Descartes, and its strict proof was demonstrated somewhat later by another mathematical genius, Leonard Euler, whose name it now carries.

Here is the complete proof of Euler's theorem, following the text of R. Courant and H. Robbins' book *What Is Mathematics?*,[3] just to show how things of that kind are done:

> To prove Euler's formula, let us imagine the given simple polyhedron to be hollow, with a surface made of thin rubber [figure 4a]. Then if we cut out one of the faces of the hollow polyhedron, we can deform the remaining surface until it stretches [figure 4b] out flat on a plane. Of course, the areas of the faces and the angles between the edges of

---

[3] The author is grateful to Drs. Courant and Robbins and to the Oxford University Press for permission to reproduce the passage that follows. Those readers who become interested in the problems of topology on the basis of the few examples given here will find a more detailed treatment of the subject in *What Is Mathematics?*

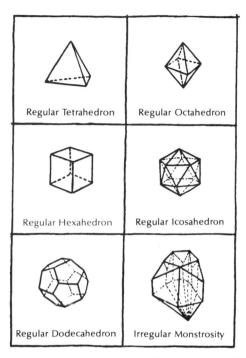

Figure 3. Five regular polyhedrons (the only possible ones) and one irregular monstrosity.

the polyhedron will be changed in this process. But the network of vertices and edges in the plane will contain the same number of vertices and edges as did the original polyhedron, while the number of polygons will be one less than in the original polyhedron since one face was removed. We shall now show that for the plane network, $V - E + F = 1$, so that, if the removed face is counted, the result is $V - E + F = 2$ for the original polyhedron.

First we "triangulate" the plane network in the following way: In some polygon of the network which is not already a triangle we draw a diagonal. The effect of this is to increase both E and F by 1, thus preserving the value of $V - E + F$. We continue now drawing diagonals, joining pairs of points until the figure consists entirely of triangles, as it must eventually [figure 4c]. In the triangulated network, $V - E + F$

plished mathematically has been to prove that five colors are always sufficient. That proof is based on the Euler relationship, which has been applied to the number of countries, the number of their boundaries, and the number of triple, quadruple, etc. points in which several countries meet.

We do not demonstrate this proof, since it is fairly complicated and would lead us away from the main subject of the discussion, but the reader can find it in various books on topology and spend a pleasant evening (and perhaps a sleepless night) in contemplating it. Either he can try to devise the proof that not only five, but even four colors are sufficient to color any map, or, if he is skeptical about the validity of this statement, he can draw a map for which four colors are not enough. In the event of success in either of the two attempts his name will be perpetuated in the annals of pure mathematics for centuries to come.

Ironically enough, the coloring problem, which so successfully eludes solution for a globe or a plane, can be solved in a comparatively simple way for more complicated surfaces such as those of a doughnut or a pretzel. For example, it has been conclusively proved that seven different colors are enough to color any possible combination of subdivisions of a doughnut without ever coloring two adjacent sections the same, and examples have been given in which the seven colors are actually necessary.

In order to get another headache the reader may get an inflated tire tube and a set of seven different paints, and try to paint the surface of the tube in such a way that each region of a given color touches six other regions of different colors. After doing it, he will be able to say that "he really knows his way around the doughnut."

Turning Space Inside Out

So far we have been discussing the topological properties of various surfaces exclusively, that is, the subspaces of only two dimensions, but it is clear that similar questions can also be asked in relation to the three-dimensional space in which we ourselves live. Thus the three-dimensional generalization of the map-coloring problem can be formulated somewhat as follows: We are asked to build a space mosaic using many variously shaped pieces of different materials, and want to do it in such a way that no two pieces made of the same material will be in contact along the common surface. How many different materials are necessary?

What is the three-dimensional analogy of the coloring problem on the surface of a sphere or torus? Can one think about some unusual three-dimensional spaces that stand in the same relation to our ordinary space, as the surfaces of the sphere or torus to the ordinary plane surface? At first the question looks senseless. In fact, whereas we can easily think of many surfaces of various shapes, we are inclined to believe that there can be only one type of three-dimensional space, namely the familiar physical space in which we live. But such an opinion represents a dangerous delusion. If we stimulate our imaginations a little, we can think of three-dimensional spaces that are rather different from that studied in the textbooks of Euclidian geometry.

The difficulty in imagining such odd spaces lies mainly in the fact that, being ourselves three-dimensional creatures, we have to look on the space so to speak "from inside," and not "from outside" as we do with various odd-shaped surfaces. But with some mental gymnastics we will conquer these odd spaces without much trouble.

Let us first try to build a model of a three-dimensional space that would have properties similar to the surface of a sphere. The main property of a spherical surface is, of course, that, though it has no boundaries, it still has a finite area; it just turns around and closes on itself. Can we imagine a three-dimensional space that would close on itself in a similar way, and thus have a finite volume without having any sharp boundaries? Think about two spherical bodies each limited by spherical surfaces, as the body of an apple is limited by its skin.

Imagine now that these two spherical bodies are put "through one another" and joined along the outer surface. Of course we do not try to tell you that one can take two physical bodies such as our two apples and squeeze them through each other so that their skins can be glued together. The apples would be squashed but would never penetrate each other.

One must rather think about an apple with an intricate system of channels eaten through it by worms. There must be *two* breeds of worm, say white and black ones, who do not like each other and never join their respective channels inside the apple although they may start them at adjacent points on the surface. An apple attacked by these two kinds of worm will finally look somewhat like figure 7, with a double network of channels, tightly intertwined and filling up the entire interior of our apple. But, although white and black channels pass very close to each other, the only way to get from one half of the labyrinth to the other is to go first through the surfaces. If you imagine the channels becoming thinner and thinner, and their number larger and larger, you will finally envisage the space inside the apple as being formed by the overlapping of two independent spaces connected only at their common surface.

If you do not like worms, you can think of a double system of enclosed corridors and stairways that could have been built, for example, inside the giant sphere at the last World's Fair in New York. Each system of stairways can be thought of as running through the entire volume of the sphere, but to get from some point of the first system to an adjacent point of the second system, one would have to go all the way to the surface of the sphere, where the two systems join, and then all the way back again. We say that two spheres overlap without interfering with each other, and a friend of yours could be very close to you in spite of the fact

Figure 7.

that in order to see him, and to shake his hand you would have to go a long way around! It is important to notice that the joining points of the two stairway systems would not actually differ from any other point within the sphere, since it would always be possible to deform the whole structure so that the joining points would be pulled inward and the points that were previously inside would come to the surface. The second important point about our model is that in spite of the fact that the total combined length of channels is finite, there are no "dead ends." You could move through the corridors and stairway on and on without being stopped by any wall or fence, and if you walked far enough you would inevitably find yourself at the point from which you started. Looking at the entire structure *from outside* one can say that a person moving through the labyrinth finally would come back to the point of his departure simply because the corridors gradually turned around, but for the people who were *inside,* and could not even know that such a thing as the "outside" existed, the space would appear as being of *finite size and yet without any marked boundaries.* . . . [T]his *"self-inclosed space of three dimensions"* that has no apparent boundaries and yet is not at all infinite was found very useful in the discussion of the properties of the universe at large. In fact, observations carried on at the very limit of telescopic power seem to indicate that at these giant distances space begins to curve, showing a pronounced tendency to come back and to close on itself in the same way as do the channels in our example of an apple eaten by the worms. But before we go on to these exciting problems, we have to learn a little more about other properties of space.

We are not yet quite through with the apple and the worms, and the next question we ask is whether it is possible to turn a worm-eaten apple into a doughnut. Oh no, we do not mean to make it taste like a doughnut, but just to make it look like one. We are discussing geometry, and not the art of cooking. Let us take a double apple such as that discussed in the previous section, that is, two fresh apples put "through one another" and "glued together" along their surfaces. Suppose a worm has eaten within one of the apples a broad circular channel as shown in figure 8. Within *one* of the apples, mind you, so that whereas outside the channel each point is a double one belonging to both apples, in-

metry, and is as we say, asymmetrical, it will be bound in two different modifications—a right- and a left-handed one. This difference occurs not only in man-made objects like gloves or golf clubs, but also very often in nature. For example, there are two varieties of snails, which are identical in all other respects, but differ in the way they build their house: one variety has the shell spiraling clockwise, whereas the other spirals in a counterclockwise way. Even the so-called molecules, the tiny particles from which all different substances are built, often possess right- and left-handed forms, very similar to those of right and left gloves, or clockwise and counterclockwise snail shells. You cannot see the molecules, of course, but the asymmetry shows up on the form of the crystals, and some optical properties of these substances. There are, for example, two different kinds of sugar, a right- and a left-handed sugar, and, believe it or not, there are also two kinds of sugar-eating bacteria, each kind consuming only the corresponding kind of sugar.

As was said above, it seems quite impossible to turn a right-handed object, a glove for example, into a left-handed one. But is that really true? Or can one imagine some tricky kind of space in which this can be done? To answer this question, let us examine it from the point of view of the flat inhabitants of a surface that can be observed by us from our superior three-dimensional outlook. Look at figure 11, representing some examples of the possible inhabitants of flatland, that is, of the space of two dimensions only. The man standing with a bunch of grapes in his hand can be called a "face-man" since he has a "face" but no "profile." The animal is, however, a "profile-donkey" or to be more specific a "right-looking-profile-donkey." Of course we can also draw a "left-looking-profile-donkey" and, since both donkeys are confined to the surface, they are just as different, from the two-dimensional point of view, as a right and a left glove in our ordinary space. You cannot superimpose a "left donkey" on a "right donkey," since in order to bring their noses and tails together you would have to turn one of them upside down, and thus his legs would be hitting the air instead of standing firmly on the ground.

But if you take one donkey out of the surface, turn it around in space, and put it back again, the two donkeys will become identical. By way of analogy one could say that a right glove can

Figure 11. An idea of two-dimensional "shadow-creatures" living on a plane. This kind of two-dimensional creature is not very "practical." The man has his face and not his profile, and cannot put into his mouth the grapes he holds in his hand. The donkey can eat the grapes all right but can walk only to the right and has to back in order to move to the left. It isn't unusual for donkeys, but not good in general.

be turned into a left glove by taking it out of our space in the fourth direction and rotating it in a proper way before putting it back. But our physical space hasn't a fourth dimension, and the above described method must be considered as quite impossible. Isn't there any other way?

Well, let us return again to our two-dimensional world, but, instead of considering an ordinary plane surface as in figure 11, investigate the properties of the so-called "surface of Möbius." This surface, named for a German mathematician who studied it first almost a century ago, can be easily made by taking a long strip of ordinary paper and gluing it into a ring, twisting it once before the two ends are joined together. Examination of figure 12 will show you how to do it. This surface has many peculiar properties, one of which can be easily discovered by cutting, with a pair of scissors, completely around it in a line parallel to the edges (along the arrows in figure 12). You would expect, of course, that by doing so, you would cut the ring into two separate rings. Do it, and you will see that your guess was wrong: instead of two

rings you will find only one ring, but one twice as long as the
original and half as wide!

Let us see now what happens to a shadow donkey when he walks
around on the Möbius surface. Suppose he starts with the position 1 (figure 12) being seen at this moment as a "left-profile donkey." On and on he goes, passing through the positions 2 and 3,
clearly visible in the picture, and finally approaches the spot from
which he started. But to your, and his, surprise, our donkey finds
itself (position 4) in an awkward position, his legs sticking up into
the air. He can, of course, turn in his surface, so that his legs will
come down, but then he will be facing the wrong way.

In short, by walking around the surface of Mobius, our "left-
profile" donkey has turned into one with a "right profile." And,
mind you, this has happened in spite of the fact that the donkey
has remained on the surface all the time and hasn't been taken up
and turned around in space. Thus we find that *on a twisted surface a right-hand object can be turned into a left-hand one, and
vice versa, by merely carrying it around the twist.* The Möbius
strip shown in figure 12 represents a part of a more general surface, known as the Klein bottle (shown on the right in figure 12),
which has only one side and closes itself, having no sharp boundaries. If this is possible on a two-dimensional surface, the same

Figure 12.   Surface of Möbius and Klein's bottle.

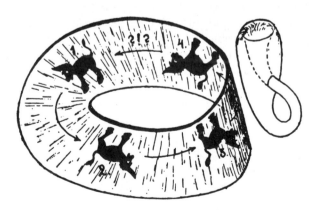

must be true also in our three-dimensional space provided of course that it is twisted in a proper way. Naturally it is not easy to imagine a Möbius twist in space. We cannot look at our space from outside, as we looked at the donkey's surface, and it is always difficult to see things clearly when you are right in the midst of them. But it isn't at all impossible that astronomical space is closed on itself and in addition twisted in the Möbius way.

If this is really so travelers around the universe would come back left-handed with their hearts in the right part of their chests, and the manufacturers of gloves and shoes would have the dubious advantage of being able to simplify the production by making only one kind of shoes and gloves, and shipping one half of them around the universe to turn them into the kind needed for the other half of the world's feet.

On this fantastic thought, we finish our discussion of the unusual properties of unusual spaces.

QUESTIONS AND SUGGESTIONS

1. Gamow's tone is thoroughly conversational. If he were talking with you, you would probably think him energetic, slightly irreverent. Compare his tone with Michael Faraday's.

2. Gamow's *Mr. Tompkins in Wonderland* is set in a world in which the speed of light is 3 mph. Relativity is observable. The reader is forced to draw comparisons with the phenomena of the real world, to think in ways he would not have thought without fiction. Similarly, above, note how many of Gamow's examples appeal to our sense of our own bodies (the one about right- and left-handed molecules, for example), or evoke familiar objects with an unfamiliar twist, like an apple eaten through by two worms. Gamow is trying to get the reader to work his mind around new concepts. Does he always succeed?

3. Summarize some of the strategies of this essay. For example, Gamow begins with a definition of space using commonsense examples, from coordinate systems; this leads to the idea of curved four-dimensional space. Then he discusses Euler's theorem. And so on.

4. Gamow's original preface claimed that his audience was a juvenile one, but he later revised this to say that perhaps the book was best suited to an older audience. Based on this selection from it, do you agree?

first three volumes were published in 1749, and the series was not complete when he died in 1788, just before the French Revolution. Buffon had aimed to produce a scientific encyclopedia dealing in succession with the planetary system, the Earth, the human race and the different kingdoms of living creatures: at the time of his death, his published volumes had got as far as the birds, and he was in the middle of cataloguing the fishes. Described so baldly, his achievement may sound pedestrian, but in fact he saw this multitudinous mass of detail in relation to a wider—and an original—scheme of ideas. Furthermore, his books contain long and penetrating digressions about scientific theory and method, and so read at times like a scientific *Tristram Shandy:* any curious and intriguing fact of nature (say, the sterility of mules) is liable to be an occasion for a far-ranging theoretical argument. Still, behind these apparent irrelevancies, there is an underlying system. His books cover some forty years of a busy career, and these theoretical asides show his point of view developing from decade to decade, in the light of new discoveries and controversies.

Buffon's conception of *Histoire Naturelle* embraced what we are here calling the history of Nature. His early hypothesis about the origin of the Solar System was intended to be the opening instalment of a story which would cover cosmological, geological and zoological development. These plans were, however, frustrated by the theologians of the Sorbonne, and he was compelled to issue a formal retraction. For the next twenty-five years Buffon kept his unorthodox speculations to himself, but by 1774 he felt sufficiently secure to return to the dangerous topics. In his *Introduction to the History of Minerals,* he described a comprehensive series of experiments on the rates at which spheres of different sizes and substances cooled down; and in a long appendix he calculated the times needed for the different planets and their satellites to cool from a white heat to an inhabitable temperature. He wrote without apology, just as though his *Theory of the Earth* had never been condemned. His purpose was concealed partly by the laboriousness of his arithmetic, partly by a mollifying heading (*partie hypothétique*). Reassured by the reception of this *Introduction* he followed it in 1778 with a fuller account of the successive *Epochs of Nature* through which the Earth had presumably developed.

The fundamental problem for the history of Nature has never been more clearly formulated than it was in his opening words:

> Just as in civil history we consult warrants, study medallions, and decipher ancient inscriptions, in order to determine the epochs of the human revolutions and fix the dates of moral events, so in natural history one must dig through the archives of the world, extract ancient relics from the bowels of the earth, gather together their fragments, and assemble again in a single body of proofs all those indications of the physical changes which can carry us back to the different Ages of Nature. This is the only way of fixing certain points in the immensity of space, and of placing a number of milestones on the eternal path of time.
>
> The past is like distance: our view of it would shrink and even be lost entirely, if history and chronology had not marked the darkest points by beacons and torches. Yet despite these lights of written tradition, let us go back a few centuries, and how uncertain are our facts! How confused are the causes of events! And what profound darkness enshrouds the periods before that tradition! Besides, it tells us only about the deeds of a few nations, i.e. the doings of a very small part of mankind: the rest of the human race is as nothing, either for us or for posterity. Thus civil history, bounded on one side by the darkness of a period not far distant from our own, embraces in the other direction only those small areas of the Earth which have been occupied in succession by peoples mindful of their own tradition. Natural history, on the other hand, embraces in its scope all regions of space and all periods equally, and has no limits other than those of the universe.

However unchanging Nature might appear to human eyes, the Earth's present state was without question very different from its original one; and, during the intervening period, it must have passed through several other phases, occupying longer or shorter times. These periods he called *epochs,* and the problem was to reconstruct them, using evidence of three different kinds:

> The surface of the Earth has taken different forms in succession; even the heavens have changed, and all the objects in the physical world are, like those of the moral world,

caught up in a continual process of successive variations . . .

But so as to pierce the night of past time—to recognize by a study of existing objects the former existence of those which have been destroyed, and work our way back to this historic truth of buried facts by the force of existing facts alone; so as to judge (in short) not merely the recent past, but also the more remote, on the basis of the present alone . . . we shall employ three prime resources: (i) those facts which can take us back to the origin of Nature; (ii) those relics which must be regarded as evidence of earlier eras; (iii) those traditions which can give us some idea about subsequent eras—after which we shall attempt to link them all together by comparisons, so as to form a single chain descending from the zenith of the scale of time down to ourselves.

By "facts," he meant those physical properties on which he founded his arguments about the cooling of the planets; by "relics," he meant such things as fossils, shells and mammoth-bones; and, by referring also to "traditions," he hoped to forestall the inevitable religious opposition—attempting to harmonize his scientific ideas with the Old Testament, by judiciously reinterpreting the opening chapters of *Genesis*. The *Book of Genesis* (he argued) had not been written for scientists, but for the unlearned. The Days of Creation could not have been "days" such as we know now—of twenty-four hours each—since the very succession of day and night was established only on the third "day," after the creation of the Sun. Rather, we should read the Biblical word "days" as referring to periods of indefinite length,.about whose exact duration Moses had not committed himself. In this way, we could rescue *Genesis* from all danger of contradicting the facts of Nature and the conclusions of reason.

With this preamble, Buffon launched into an account of the geological epochs corresponding to the seven Days of Creation. These epochs had lengths ranging from 3000 to 35,000 years: in all, the Earth's history had by now occupied some 75,000 years, and a further 93,000 remained before life would be extinguished by cold. For the first epoch, he revived his theory about a near-collision between the Sun and a passing comet; in the second, he supposed the Earth to have solidified, with the greater part of the fusible rocks on the surface; during the third, all the continents were covered with water; fourthly, the oceans withdrew and volcanoes built up the land; during the fifth epoch, tropical animals were spread across the whole Earth; in the sixth, the different continents separated; and finally, in the seventh and last epoch, one reached the period of Man's existence.

Taken separately, none of Buffon's chief steps was entirely new. Descartes had spoken of the planets as having formed out of incandescent stars, Leibniz had developed this idea further in his *Protogea*, and even Newton had used the same idea to explain the flattening of the Earth at the Poles. (This was the result of centrifugal action while the planet was still hot and plastic.) This idea of interpreting the Days of Creation as geological epochs had been hinted at in the 1690s, during the discussion prompted by Burnet's *Sacred History of the Earth*. Again, Benoit de Maillet's posthumous dialogue, *Telliamed*, had foreshadowed Buffon's extension of the time-scale: the "Indian sage" into whose mouth de Maillet put his own original speculations had insisted that we should not "fix a beginning to that which perhaps never had one. Let us not measure the past duration of the world by that of our own years." Finally, Buffon's experiments on cooling were consciously modelled on those by which Newton had established the basic laws governing the rate at which bodies lose heat. In fact, Newton had gone so far as to enquire, in an incidental note, how long it would take for an iron sphere the size of the Earth to cool down from red heat, and had arrived at an answer comparable to Buffon's—*viz.*: 50,000 years. For once, Newton's calculations led to a result that he could not square with his religious convictions, and he unhesitatingly rejected it: there must be something wrong with his calculation—perhaps his assumption that a large sphere would cool very much more slowly than a small one—and he added: "I should be glad that the true ratio was investigated by experiments."

What distinguished Buffon's *Epochs of Nature* was the cumulative weight of his whole argument. He drew together into a unified whole half a dozen ideas which had previously been thrown out independently. Furthermore, he patiently settled down to work out, in numerical terms, the actual periods of time demanded by a physical theory of the Earth's development. To summarize his 1774 figures for the Earth's cooling:

(i) Organized nature as we know it, is not yet born on Jupiter, whose heat is still too great today for one to touch its surface, and it will only be in 40,791 years [i.e. 115,623–74,832: cf. table above] that living creatures will be able to subsist there, but thereafter once established they would last 367,498 years on that large planet;

(ii) living nature, as we know it, has been extinct on the fifth satellite of Saturn for the last 27,274 years; on Mars, for the last 14,506 years, and on the Moon for the last 2,318 years; [etc. etc.] . . .

[Hence my belief in] the real existence of organized and sensible beings on all the bodies of the solar system, and the more-than-likely existence of the same beings on all the other bodies making up the systems of other suns, so augmenting and multiplying almost to infinity the extent of living Nature, and at the same time raising the greatest of all monuments to the glory of the Creator.

By our standards, of course, Buffon's calculations had given the Earth not too long, but far too short a life. Where did his physics go astray? He acknowledged that the present temperature of any planet or satellite must be determined by two separate factors—the radiant heat falling on it from the Sun, and the residual heat remaining from its original molten state—and he did his best to estimate the relative contributions of these two factors. In the 1770s the laws of normal cooling were well established, but radiant heat was little understood: Buffon mistakenly decided that the effects of solar radiation were of secondary importance as compared with the Earth's residual heat. He would have reached a more accurate result if he had assumed the exact reverse. The present temperature of a planet or satellite depends almost entirely on the amount of incoming solar radiation: to a first approximation, one can neglect the heat coming from the interior of the planet itself. This fact became evident soon after 1800, when physicists began to study radiant heat seriously, and within forty years of their original publication Buffon's calculations were completely undercut by the new ones of Fourier.

Scientifically, then, Buffon's pioneer attempt to estimate the age of the Earth by appeal to physical principles had been a failure. His religious compromise, equally, had satisfied none of the parties affected. So, in retrospect, the *Epochs of Nature* may seem at best a

magnificent ruin. Yet, this verdict would be unjust. Buffon's figures were wrong; but his was the voice of the future, and—above
all—his books were very widely read. Moreover, his calculations
had proved the essential point: that the time-barrier could be
breached. By invoking the laws governing familiar physical processes, such as cooling, one might infer the former state of things
from the present face of Nature, and determine the dates of physical events far earlier than the first human records. If Buffon had
lived to see the next fifty years of geology, he would have been well
content, for by the 1820s literal-minded fundamentalism was in
full retreat, and the defenders of orthodoxy were thankful to take
refuge in his own interpretation of the Days of Creation. The details of his argument could go overboard: he had made the points
that mattered.

The Fact of Geological Change

As things turned out, Buffon's own theories about the age of the
Earth made little immediate contribution to geology proper, and
the cosmological approach to the subject soon went out of fashion. For his was a speculative and roundabout way of arguing, in
which the history of the Earth was deduced indirectly from a general theory of the planetary system, rather than being pieced together directly from the actual evidence of geological exploration.
The results might carry a certain abstract conviction to Newtonian
natural philosophers, but they had no great relevance to the experience of practical men, and they were not easily reconciled with
the ideas such men inherited from their predecessors. Indeed, as
we shall see later, until well into the twentieth century certain
glaring discrepancies remained between physical cosmology on the
one hand, and zoology and geology on the other.

Meanwhile, however, other men were scrutinizing the Earth's
surface directly, and beginning to enquire about the agencies that
had shaped it into its present form. At first—as we said—they were
not moved by historical curiosity: they were concerned merely to
map the rock-strata found in the Earth's crust at different places,
and to discover whether there was any common sequence in the
formations overlying one another in different countries and re-

gions. It was some time before the discovery of a widespread *geographical* order in the nature of the crust came to be recognized as evidence of a *temporal* order in the processes by which the rocks had come into existence—instead of being accepted unquestioningly as the pattern imposed at the original Creation. And it took most of a century for geologists to establish what agencies had been involved in these processes of formation, to discover how one might compare the ages of strata geographically distant from one another, and to build up a consistent history of the Earth's crust from the evidence of its present form and fossil content.

To begin with, there were two chief centres of geological research, one in Germany, the other in France. Ever since mediaeval times, German craftsmen had been building up a tradition of mineralogy and mining, and this practical aspect marked the first German contributions to the new science. Anyone with first-hand experience of mining technique knew something about the stratification of rock-layers, and much was done to lay the foundations for geology by simply describing and naming the different types of rock and rock-strata. The chief leader in this work was the Saxon geologist, A. G. Werner. Like Boerhaave, Werner was one of the supreme scientific teachers, who made his mark primarily by stimulating the interest of his students; and as with that other great teacher, Linnaeus, the men who learned from him were won over by his intimate and detailed mastery over his subject-matter, rather than because of his theoretical penetration. In geology as much as in zoology and botany, what the eighteenth century demanded was a comprehensive and orderly classification, together with a precise nomenclature. These Werner provided: he classified the superimposed rock-strata into four or five main types, and several dozen subdivisions—ranging from granite and gneiss, in which fossils were never found, up through the various fossil-bearing strata to the sands, clays and volcanic lava of the surface layers—and he saw that the superposition of layers must have a historical significance also, the fossil-bearing rocks being younger than the granite, and the superficial rocks being the most recent of all.

In France, meanwhile, other historical clues were coming to light. In 1751, J. E. Guettard stopped at Montélimar on his way back from a journey throughout southern Italy. His attention was caught by the fact that the paving of the streets was a type of stone

—hexagonal basalt—strikingly like some which he had seen in the volcanic regions around Vesuvius, and that the milestones and even some of the local buildings were also made of volcanic-looking stone. He enquired where this stone had come from, and was directed into the mountains west of the Rhône. Here he found a region of steep river valleys, containing dramatic cliffs of basalt "organ-pipes," and leading to a central plateau dominated by a range of conical peaks. One of these mountains was the Puy de Dôme, the site of Pascal's experiments on atmospheric pressure. In the neighbouring village of Volvic, men had been quarrying a durable black stone for centuries. It had been used to build the cathedral at Clermont Ferrand, and is still used for milestones on French roads to this day. Until Guettard arrived, however, the quarries of Volvic and the neighbouring peaks had kept their most important secret, which could be discerned only by men who looked at them with the right questions in mind. As Guettard immediately realized, the surrounding mountain peaks were the cones of extinct volcanoes, and the paths of their ancient lava-streams could be clearly seen in the surrounding countryside. One had only to strip off the surface disguise of top-soil and vegetation, and the whole region was recognizable as a vast area shaped by volcanic action.

Like the first deciphering of a hitherto-unintelligible script, Guettard's discovery precipitated a chain of others. Once the first step was taken, everything else fell neatly into place. During the decades that followed, similar regions of extinct volcanoes were recognized in a dozen parts of the world, many of them associated with the hexagonal basalt characteristic of Giants' Causeways. Yet the puzzle remained: if the time-span of the world was really less than 6000 years, how could such violent volcanic action have gone on unrecorded? It seemed impossible that all these formidable eruptions could have taken place during the short span of time before the earliest human records. Faced with evidence such as this men were gradually driven towards the conclusion which they had so long resisted: that the present phase in the Earth's history was only the most recent in a series of prolonged epochs, and that the face of the globe had changed radically from age to age.

QUESTIONS AND SUGGESTIONS

1. Toulmin and Goodfield tell the story of a great event in the history of Western thought: the breaking of the "time barrier," the extension of man's consciousness of his own past far beyond the limits of recorded history. If with the invention of the telescope man looked into an infinitely greater space than he had imagined before, so with the development of geology man's imagination reached back across aeons to a time before he existed. By what methods do we reconstruct the past, whether we are historians, archeologists, geologists, or evolutionary biologists? How are bits of evidence used to infer some aspect of what the world was like? Much of the information you receive is about the past. How do you evaluate it?

2. The development of modern methods of historical study and research were a precondition for the earth's acquisition of a history. What similarities exist between the methods of the historian and those of the historical geologist?

3. If the earth had always existed in the same state, would we be able to know that it had a history?

4. Buffon was wrong in many of his calculations, but they challenged existing opinion in the right way. Toulmin and Goodfield aim to put geology into the context of cultural history; Buffon is important in illustrating a climate of ideas. Write an essay on the relationship of some aspect of contemporary scientific research to our cultural "climate of opinion": for example, space research, genetic engineering, agriculture, brain research.

5. Fractal research involves the mathematical study of shapes, some of which are geographic. Look into Bernard Mandelbrot's book *The Fractal Geometry of Nature* (New York: W. H. Freeman, 1983) or his essay "How Long Is the Coast of Britain?" in *Science* 155 (1967): 636–638 for a report on the place of fractal geometry in the study of earth science.

KEITH TINKLER

What Is Geomorphology?

KEITH TINKLER (b. 1942) is professor of geography at Brock University in St. Catherines, Ontario. He is the author of more than forty articles on geographical and geological subjects and his work has appeared in *Professional Geographer, The Canadian Geographer, The Bulletin of the Geological Society of America, Nature,* and *Area.* He is especially interested in the Niagara Falls region and postglacial lake formation. Dr. Tinkler is the author of *Nysteun-Dacey Nodal Analysis* (1988) and the editor of *History of Geomorphology: From Hutton to Hack* (1989). The following selection is from his first book, *A Short History of Geomorphology* (1985).

Geomorphology, as currently construed, is concerned with the shape of the earth's surface, more or less as we see it, and the processes that are involved in changing this shape over time. The subject falls most naturally within the province of geology, although in much of the English-speaking world, and especially the Commonwealth and former British Empire countries, geomorphology is institutionally associated with geography rather than geology as it is in the United States.

The global aspects of earth (geo . . .) shape (. . . morph . . .) science (. . . logos) implied in the word itself are not usually included in its study. The study of earth shape taken at the global scale is the province of geophysics, in which subject it was already securely lodged by the time the word *geomorphology* came into general use after about 1890. There is, nevertheless, an interaction between the concerns of geophysics and geomorphology at the regional and the continental scale, where land and sea beds are deformed by the varying loads placed upon them by water, ice, and sediment and their corollaries, the melting of ice, the erosion of sediment from the continents, and the changing sea levels of the last two million years. It was not until well into the present century that the importance of these interactions was understood properly,

95

potentially it might have been much larger than it was. The construction of a "regime" canal is a far cry from a natural river. The canal must transport water, but as far as possible not sediment, and so, when constructed from local natural materials, a very delicate balance must be sought. In contrast, geologists have sought to understand how rivers are able to cut and maintain their channels, adjust to rapidly changing bed and bank materials, and still transport a varied load of sediment. Despite their opposing aims, both studies have much in common and it is all the more surprising that eighteenth- and nineteenth-century engineers in France and Italy had worked out the basic hydraulics of canals and devised the notion of the profile of equilibrium at a time when many geologists denied that valleys were eroded by rivers, or even that they transported sediment to the sea (de Luc 1790, Kirwan 1799). By the time fluvial erosion of valleys was generally accepted, after about 1862, interest in geomorphology concerned landform evolution at the regional scale and over millions of years.

After World War II, a serious interest in the geomorphology of flooding in the United States led to a revitalization of the discipline by the infusion of ideas from civil engineering. Currently much work is being done at the interfaces of geomorphology, hydrology, and hydraulic engineering and, although the main emphasis in recent decades has been upon processes at short timescales of the order of one to one hundred years, longer timescales are coming into focus once more.

QUESTIONS AND SUGGESTIONS

1. What is the dominant style of this excerpt from the first chapter of Tinkler's book *A Short History of Geomorphology*?
2. What comparisons can be drawn between the styles of Toulmin and Goodfield, Gould and Tinkler?
3. Research and write on the geomorphological history of a region in a state or a country that interests you.
4. Locate definitions of geography, geomorphology, and geology in specialized dictionaries or encyclopedias. Prepare a comparative analysis.

STEPHEN J. GOULD

False Premise, Good Science

STEPHEN J. GOULD (b. 1941) has been a professor of geology at Harvard since 1973, where he is also the curator of Invertebrate Paleontology for Harvard's Museum of Comparative Zoology. Trained as a paleontologist, Gould received his Ph.D. from Columbia for his work on fossil land snails in Bermuda. Since 1974 he has contributed a monthly essay to *Natural History*, a publication of the American Museum of Natural History in New York City. A noted author of popular scientific essays and books, Gould received both the National Book Critics' Circle Award for *The Mismeasure of Man* and the American Book Award for Science for *The Panda's Thumb* in 1981. Recurrent themes in Gould's writings include Darwin's biological theories, the cultural bias of Americans against science, the influence of society on scientific theory, and sociobiology. Gould's engaging style and clarity of expression have been compared to Lewis Thomas's writing on science. The selection reprinted here is from Gould's 1985 book, *The Flamingo's Smile: Reflections on Natural History*.

My vote for the most arrogant of all scientific titles goes without hesitation to a famous paper written in 1866 by Lord Kelvin, "The 'Doctrine of Uniformity' in Geology Briefly Refuted." In it, Britain's greatest physicist claimed that he had destroyed the foundation of an entire profession not his own. Kelvin wrote:

> The "Doctrine of Uniformity" in Geology, as held by many of the most eminent of British geologists, assumes that the earth's surface and upper crust have been nearly as they are at present in temperature and other physical qualities during millions of millions of years. But the heat which we know, by observation, to be now conducted out of the earth yearly is so great, that if *this* action had been going on with any approach to uniformity for 20,000 million years, the amount of heat lost out of the earth would have been about as much as would heat, by 100° Cent., a quantity of ordinary surface rock of 100 times the earth's bulk. (See calculation appended.) This would be more than enough to melt a mass of surface rock equal in bulk to the *whole earth*. No hypothesis as to chemical action, internal fluidity, effects of

> pressure at great depth, or possible character of substances in the interior of the earth, possessing the smallest vestige of probability, can justify the supposition that the earth's crust has remained nearly as it is, while from the whole, or from any part, of the earth, so great a quantity of heat has been lost.

I apologize for inflicting so long a quote so early in the essay, but this is not an extract from Kelvin's paper. It is the whole thing (minus the appended calculation). In a mere paragraph, Kelvin felt he had thoroughly undermined the very basis of his sister discipline.

Kelvin's arrogance was so extreme, and his later comeuppance so spectacular, that the tale of his 1866 paper, and of his entire, relentless forty-year campaign for a young earth, has become the classical moral homily of our geological textbooks. But beware of conventional moral homilies. Their probability of accuracy is about equal to the chance that George Washington really scaled that silver dollar clear across the Rappahannock.

The story, as usually told, goes something like this. Geology, for several centuries, had languished under the thrall of Archbishop Ussher and his biblical chronology of but a few thousand years for the earth's age. This restriction of time led to the unscientific doctrine of catastrophism—the notion that miraculous upheavals and paroxysms must characterize our earth's history if its entire geological story must be compressed into the Mosaic chronology. After long struggle, Hutton and Lyell won the day for science with their alternative idea of uniformitarianism—the claim that current rates of change, extrapolated over limitless time, can explain all our history from a scientific standpoint by direct observation of present processes and their results. Uniformity, so the story goes, rests on two propositions: essentially unlimited time (so that slow processes can achieve their accumulated effect), and an earth that does not alter its basic form and style of change throughout this vast time. Uniformity in geology led to evolution in biology and the scientific revolution spread. If we deny uniformity, the homily continues, we undermine science itself and plunge geology back into its own dark ages.

Yet Kelvin, perhaps unaware, attempted to undo this triumph

of scientific geology. Arguing that the earth began as a molten body, and basing his calculation upon loss of heat from the earth's interior (as measured, for example, in mines), Kelvin recognized that the earth's solid surface could not be very old—probably 100 million years, and 400 million at most (although he later revised the estimate downward, possibly to only 20 million years). With so little time to harbor all of evolution—not to mention the physical history of solid rocks—what recourse did geology have except to its discredited idea of catastrophes? Kelvin had plunged geology into an inextricable dilemma while clothing it with all the prestige of quantitative physics, queen of the sciences. One popular geological textbook writes (C. W. Barnes), for example:

> Geologic time, freed from the constraints of literal biblical interpretation, had become unlimited; the concepts of uniform change first suggested by Hutton now embraced the concept of the origin and evolution of life. Kelvin single-handedly destroyed, for a time, uniformitarian and evolutionary thought. Geologic time was still restricted because the laws of physics bound as tightly as biblical literalism ever had.

Fortunately for a scientific geology, Kelvin's argument rested on a false premise—the assumption that the earth's current heat is a residue of its original molten state and not a quantity constantly renewed. For if the earth continues to generate heat, then the current rate of loss cannot be used to infer an ancient condition. In fact, unbeknown to Kelvin, most of the earth's internal heat is newly generated by the process of radioactive decay. However elegant his calculations, they were based on a false premise, and Kelvin's argument collapsed with the discovery of radioactivity early in our century. Geologists should have trusted their own intuitions from the start and not bowed before the false lure of physics. In any case, uniformity finally won and scientific geology was restored. This transient episode teaches us that we must trust the careful empirical data of a profession and not rely too heavily on theoretical interventions from outside, whatever their apparent credentials.

So much for the heroic mythology. The actual story is by no means so simple or as easily given an evident moral interpretation.

based his argument upon so many unprovable assumptions (about the earth's uniform composition, for example) that he could not calculate a precise figure for the earth's age.

Thus, although all three arguments had a quantitative patina, none was precise. All depended upon simplifying assumptions that Kelvin could not justify. All therefore yielded only vague estimates with large margins of error. During most of Kelvin's forty-year campaign, he usually cited a figure of 100 million years for the earth's age—plenty of time, as it turned out, to satisfy nearly all geologists and biologists.

Darwin's strenuous opposition to Kelvin is well recorded, and later commentators have assumed that he spoke for a troubled consensus. In fact, Darwin's antipathy to Kelvin was idiosyncratic and based on the strong personal commitment to gradualism so characteristic of his world view. So wedded was Darwin to the virtual necessity of unlimited time as a prerequisite for evolution by natural selection that he invited readers to abandon *The Origin of Species* if they could not accept this premise: "He who can read Sir Charles Lyell's grand work on the *Principles of Geology,* and yet does not admit how incomprehensively vast have been the past periods of time, may at once close this volume." Here Darwin commits a fallacy of reasoning—the confusion of gradualism with natural selection—that characterized all his work and that inspired Huxley's major criticism of the *Origin:* "You load yourself with an unnecessary difficulty in adopting *Natura non facit saltum* [Nature does not proceed by leaps] so unreservedly." Still, Darwin cannot be entirely blamed, for Kelvin made the same error in arguing explicitly that his young age for the earth cast grave doubt upon natural selection as an evolutionary mechanism (while not arguing against evolution itself). Kelvin wrote:

> The limitations of geological periods, imposed by physical science, cannot, of course, disprove the hypothesis of transmutation of species; but it does seem sufficient to disprove the doctrine that transmutation has taken place through "descent with modification by natural selection."

Thus, Darwin continued to regard Kelvin's calculation of the earth's age as perhaps the gravest objection to his theory. He wrote to Wallace in 1869 that "Thomson's [Lord Kelvin's] views on the recent age of the world have been for some time one of my sorest troubles." And, in 1871, in striking metaphor, "But then comes Sir W. Thomson like an odious spectre." Although Darwin generally stuck to his guns and felt in his heart of hearts that something must be wrong with Kelvin's calculations, he did finally compromise in the last edition of the *Origin* (1872), writing that more rapid changes on the early earth would have accelerated the pace of evolution, perhaps permitting all the changes we observe in Kelvin's limited time:

> It is, however, probable, as Sir William Thompson [*sic*] insists, that the world at a very early period was subjected to more rapid and violent changes in its physical conditions than those now occurring; and such changes would have tended to induce changes at a corresponding rate in the organisms which then existed.

Darwin's distress was not shared by his two leading supporters in England, Wallace and Huxley. Wallace did not tie the action of natural selection to Darwin's glacially slow time scale; he simply argued that if Kelvin limited the earth to 100 million years, then natural selection must operate at generally higher rates than we had previously imagined. "It is within that time [Kelvin's 100 million years], therefore, that the whole series of geological changes, the origin and development of all forms of life, must be compressed." In 1870, Wallace even proclaimed his happiness with a time scale of but 24 million years since the inception of our fossil record in the Cambrian explosion.

Huxley was even less troubled, especially since he had long argued that evolution might occur by saltation, as well as by slow natural selection. Huxley maintained that our conviction about the slothfulness of evolutionary change had been based on false and circular logic in the first place. We have no independent evidence for regarding evolution as slow; this impression was only an inference based on the assumed vast duration of fossil strata. If Kelvin now tells us that these strata were deposited in far less time, then our estimate of evolutionary rate must be revised correspondingly.

Biology takes her time from geology. The only reason we have for believing in the slow rate of the change in living forms is the fact that they persist through a series of deposits which, geology informs us, have taken a long while to make. If the geological clock is wrong, all the naturalist will have to do is to modify his notions of the rapidity of change accordingly.

Britain's leading geologists tended to follow Wallace and Huxley rather than Darwin. They stated that Kelvin had performed a service for geology in challenging the virtual eternity of Lyell's world and in "restraining the reckless drafts" that geologists so rashly make on the "bank of time," in T. C. Chamberlin's apt metaphor. Only late in his campaign, when Kelvin began to restrict his estimate from a vague and comfortable 100 million years (or perhaps a good deal more) to a more rigidly circumscribed 20 million years or so did geologists finally rebel. A. Geikie, who had been a staunch supporter of Kelvin, then wrote:

> Geologists have not been slow to admit that they were in error in assuming that they had an eternity of past time for the evolution of the earth's history. They have frankly acknowledged the validity of the physical arguments which go to place more or less definite limits to the antiquity of the earth. They were, on the whole, disposed to acquiesce in the allowance of 100 millions of years granted them by Lord Kelvin, for the transaction of the long cycles of geological history. But the physicists have been insatiable and inexorable. As remorseless as Lear's daughters, they have cut down their grant of years by successive slices, until some of them have brought the number to something less than ten millions. In vain have geologists protested that there must be somewhere a flaw in a line of argument which tends to results so entirely at variance with the strong evidence for a higher antiquity.

Kelvin's Scientific Challenge and the Multiple Meanings of Uniformity

As a master of rhetoric, Charles Lyell did charge that anyone who challenged his uniformity might herald a reaction that would send geology back to its prescientific age of catastrophes. One meaning of uniformity did uphold the integrity of science in this sense—the claim that nature's laws are constant in space and time, and that miraculous intervention to suspend these laws cannot be permitted as an agent of geological change. But uniformity, in this methodological meaning, was no longer an issue in Kelvin's time, or even (at least in scientific circles) when Lyell first published his *Principles of Geology* in 1830. The scientific catastrophists were not miracle mongers, but men who fully accepted the uniformity of natural law and sought to render earth history as a tale of *natural* calamities occurring infrequently on an ancient earth.

But uniformity also had a more restricted, substantive meaning for Lyell. He also used the term for a particular theory of earth history based on two questionable postulates: first, that rates of change did not vary much throughout time and that slow and current processes could therefore account for all geological phenomena in their accumulated impact; second, that the earth had always been about the same, and that its history had no direction, but represented a steady state of dynamically constant conditions.

Lyell, probably unconsciously, then performed a clever and invalid trick of argument. Uniformity had two distinct meanings—a methodological postulate about uniform laws, which all scientists had to accept in order to practice their profession, and a substantive claim of dubious validity about the actual history of the earth. By calling them both uniformity, and by showing that all scientists were uniformitarians in the first sense, Lyell also cleverly implied that, to be a scientist, one had to accept uniformity in its substantive meaning as well. Thus, the myth developed that any opposition to uniformity could only be a rearguard action against science itself—and the impression arose that if Kelvin was attacking the "doctrine of uniformity" in geology, he must represent the forces of reaction.

In fact, Kelvin fully accepted the uniformity of law and even based his calculations about heat loss upon it. He directed his attack against uniformity only upon the substantive (and dubious) side of Lyell's vision. Kelvin advanced two complaints about this substantive meaning of uniformity. First, on the question of rates. If the earth were substantially younger than Lyell and the strict uniformitarians believed, then modern, slow rates of change would not be sufficient to render its history. Early in its history, when the earth was hotter, causes must have been more energetic and intense. (This is the "compromise" position that Darwin finally adopted to explain faster rates of change early in the history of life.) Second, on the question of direction. If the earth began as a molten sphere and lost heat continually through time, then its history had a definite pattern and path of change. The earth had not been perennially the same, merely changing the position of its lands and seas in a never-ending dance leading nowhere. Its history followed a definite road, from a hot, energetic sphere to a cold, listless world that, eventually, would sustain life no longer. Kelvin fought, within a scientific context, for a *short-term, directional* history against Lyell's vision of an essentially eternal steady-state. Our current view represents the triumph of neither vision, but a creative synthesis of both. Kelvin was both as right and as wrong as Lyell.

Radioactivity and Kelvin's Downfall

Kelvin was surely correct in labeling as extreme Lyell's vision of an earth in steady-state, going nowhere over untold ages. Yet, our modern time scale stands closer to Lyell's concept of no appreciable limit than to Kelvin's 100 million years and its consequent constraint on rates of change. The earth is 4.5 billion years old.

Lyell won this round of a complicated battle because Kelvin's argument contained a fatal flaw. In this respect, the story as conventionally told has validity. Kelvin's argument was not an inevitable and mathematically necessary set of claims. It rested upon a crucial and untested assumption that underlay all Kelvin's calculations. Kelvin's figures for heat loss could measure the earth's age only if that heat represented an original quantity gradually dissipated through time—a clock ticking at a steady rate from its initial reservoir until its final exhaustion. But suppose that new heat is constantly created and that its current radiation from the earth reflects no original quantity, but a modern process of generation. Heat then ceases to be a gauge of age.

Kelvin recognized the contingent nature of his calculations, but the physics of his day included no force capable of generating new heat, and he therefore felt secure in his assumption. Early in his campaign, in calculating the sun's age, he had admitted his crucial dependence upon no new source of energy, for he had declared his results valid "unless new sources now unknown to us are prepared in the great storehouse of creation."

Then, in 1903, Pierre Curie announced that radium salts constantly release newly generated heat. The unknown source had been discovered. Early students of radioactivity quickly recognized that most of the earth's heat must be continually generated by radioactive decay, not merely dissipating from an originally molten condition—and they realized that Kelvin's argument had collapsed. In 1904, Ernest Rutherford gave this account of a lecture, given in Lord Kelvin's presence, and heralding the downfall of Kelvin's forty-year campaign for a young earth:

> I came into the room, which was half dark, and presently spotted Lord Kelvin in the audience and realized that I was in for trouble at the last part of the speech dealing with the age of the earth, where my views conflicted with his. To my relief, Kelvin fell fast asleep, but as I came to the important point, I saw the old bird sit up, open an eye and cock a baleful glance at me! Then a sudden inspiration came, and I said Lord Kelvin had limited the age of the earth, provided no new source of heat was discovered. That prophetic utterance refers to what we are now considering tonight, radium!

Thus, Kelvin lived into the new age of radioactivity. He never admitted his error or published any retraction, but he privately conceded that the discovery of radium had invalidated some of his assumptions.

The discovery of radioactivity highlights a delicious double irony. Not only did radioactivity supply a new source of heat that destroyed Kelvin's argument; it also provided the clock that could then measure the earth's age and proclaim it ancient after all! For radioactive atoms decay at a constant rate, and their dissipation does measure the duration of time. Less than ten years after the discovery of radium's newly generated heat, the first calculations for radioactive decay were already giving ages in billions of years for some of the earth's oldest rocks.

We sometimes suppose that the history of science is a simple story of progress, proceeding inexorably by objective accumulation of better and better data. Such a view underlies the moral homilies that build our usual account of the advance of science—for Kelvin, in this context, clearly impeded progress with a false assumption. We should not be beguiled by such comforting and inadequate stories. Kelvin proceeded by using the best science of his day, and colleagues accepted his calculations. We cannot blame him for not knowing that a new source of heat would be discovered. The framework of his time included no such force. Just as Maupertuis lacked a proper metaphor for recognizing that embryos might contain coded instructions rather than preformed parts, Kelvin's physics contained no context for a new source of heat.

The progress of science requires more than new data; it needs novel frameworks and contexts. And where do these fundamentally new views of the world arise? They are not simply discovered by pure observation; they require new modes of thought. And where can we find them, if old modes do not even include the right metaphors? The nature of true genius must lie in the elusive capacity to construct these new modes from apparent darkness. The basic chanciness and unpredictability of science must also reside in the inherent difficulty of such a task.

QUESTIONS AND SUGGESTIONS

1. What stylistic techniques does Gould employ to engage his readers in a discussion of Kelvin's geological theories? How would you describe Gould's audience?

2. Select your own scientific theory and write an article similar to Gould's.

3. Develop an essay on the idea that false premises can lead to good conclusions.

4. Write a theoretical essay on the idea of "progress" in science.

5. Develop a longer essay on a topic you select from reading the Gould, Toulmin and Goodfield, and Tinkler essays.

HOWARD ENSIGN EVANS

In Defense of Magic: The Story of Fireflies

HOWARD ENSIGN EVANS (b. 1919) has been a professor of entomology since 1949 and curator at the Museum of Comparative Biology at Harvard since 1964. He writes frequently about the insect world for popular audiences. "In Defense of Magic" is from *Life on a Little Known Planet,* a collection of essays about insects. Evans's essays appear in *The New Yorker,* whose standards for prose are high and which has for many years appealed to a highly educated, cosmopolitan audience.

If magic be defined as something "produced by secret forces in nature," and "secret" in turn defined as something "revealed to none or to few" (and these are legitimate definitions), then magic is not likely to be diminished by all the science we can muster. Research may provide us with answers, but these answers forever lead to new and more profound questions; and as our knowledge of the world grows more and more vast, most phenomena that can be said to be "revealed" will of necessity be revealed to fewer and fewer of us. True, the universe is orderly (more or less) and finite (probably), so it is possible to imagine a time when all knowledge has been stored in a system of computers from which any fact or combination of facts can be elicited by pushing a button. As I say, it is possible to imagine this. But just recently I have been reading several articles on animal behavior written between 1900 and 1920. Their language seemed a bit naïve by modern standards; but I wondered if we had really advanced very much in this field in half a century. Recently, too, I have been studying a group of small wasps known as the Bethylidae, a group known to only a few persons and in which probably more than half the species are, in fact, unknown to anyone—and we don't really "know" any of the several hundred "known" species in any real sense, although a few of them are reported to have rather remarkable behavior patterns.

True, some of my colleagues are deciphering the genetic code. But how much can they tell us about the production of a bethylid wasp from a particular series of chemical messages? And why, of all possible concatenations of atoms, are there wasps at all?

Magic in the sense of something "inciting wonder" is also here to stay; or if it is not, man will have been vastly diminished by its loss. One need not be standing silent on a peak in Darien. There is magic, indeed, in the crash of surf on an unknown shore; but so there is in a mud puddle. There is a powerful magic in a crash of thunder, even more powerful in a nuclear explosion; but there is a very special magic in a child's kite or in the call of a gull and all that it evokes. Mark that leaf blown before the wind: it is important; no matter how sophisticated or blasé we become, that moment, this experience, is all the treasure we shall reap in our few moments of identity.

What can rival a twilit meadow rich with the essence of June and spangled with fireflies? Here is magic, indeed, and the joy of pursuing through grass just touched with early dew a light now here, now there, now gone. Or of collecting several in a bottle and taking them indoors for illumination; or of tying one lightly with a thread to one's clothing, as natives of some tropical countries are reported to do at fiesta time. As children, we used to call them lightning bugs; and wingless kinds that emit a steady light from the ground are called glowworms in English-speaking countries wherever they occur. In fact fireflies are neither flies nor bugs nor worms, but soft-bodied beetles called Lampyridae, a name based on an old Greek word that also evolved into our word "lamp."

Some of our commonest Lampyridae, curiously, give no light at all; these are day-flying beetles that one often finds on tree trunks, looking very much like ordinary fireflies but lacking the whitish "lamps" in their tails. The common European glowworm is a wingless female that produces a steady light, while the male of the same species is winged and not luminescent. Most fireflies of eastern North America are winged, and produce a flashing light in both sexes. The larvae (and even the eggs!) of many fireflies also glow. This seems strange when we consider that the lights of fireflies are used by the adults to find the opposite sex of their own species in the dark. What function does luminescence serve in the eggs and larvae? One might assume that the immature stages simply "can't help glowing," since the rudiments of the light organs are developing within them. But the fact is that the larval and adult organs are of quite different nature, and if the larval light-producing cells are carefully excised, the adult will still develop normal light organs.

Luminescence probably first arose as a dim and diffuse product of certain normal body processes, for many substances oxidized slowly in the dark produce a glow, and a dim luminescence occurs in many simple organisms (especially in the sea). Natural light is known to occur in certain bacteria, fungi, one-celled animals, sponges, jellyfish-like animals, corals, marine worms, clams, snails, squids, arthropods, and of course a variety of deep-sea fishes—but never among the reptiles, birds, or mammals. It is possible that the earliest organisms on earth lived in an atmosphere devoid of oxygen. When oxygen first appeared—from the effects of sunlight on water vapor or from photosynthesis by primitive plants—it may have been toxic to these organisms. Luminescence may have developed as a system of getting rid of oxygen by burning it off as a "cold light." Later on, when plants and animals evolved that took advantage of oxygen to run their own body machinery, luminescence was preserved in a wide variety of organisms simply as a hangover from these ancient times. At least such is the belief of William McElroy and Howard Seliger, of Johns Hopkins University, our current leading authorities in this field. Their theory is supported by the fact that in many simple organisms luminescence seems to serve no function, and in fact in some cases a single species exists in both luminous and nonluminous forms, both apparently successful. They also point out that luminescence requires oxygen in only very low concentrations, as it must have once occurred on earth. Certain bacteria, for example, produce light when the oxygen concentration is as low as one part in 100 million.

Obviously, some of the more complex animals—fish and insects, for instance—have elaborated this primitive light-producing capacity into specialized organs serving important functions in their lives. Adult fireflies possess the most complex light organs known, and these organs are still far from fully understood. Despite the intensity of the light they produce, the amount of heat is negligible. Only in very recent years has man developed chemical light-producing systems that rival that of the firefly in efficiency.

E. Newton Harvey, of Princeton University, has written a fascinating account of the history of human knowledge of luminescence. According to Professor Harvey, the firefly is not mentioned in the Bible, the Talmud, or the Koran, probably because fireflies are absent or uncommon in the arid regions of the Near East. However, the Chinese *Book of Odes,* dating from 1500 to 1000 B.C., speaks of the "fitful light of glowworms," and there are many accounts of fireflies in ancient writings of the Far East. The Japanese believed fireflies to be transformed from decaying grasses, while glowworms were said to arise from bamboo roots. In Japan, firefly collecting was popular in early times, and there is said to have been a firefly festival each year near Kyoto.

Aristotle was familiar with fireflies, and was apparently aware that some glowworms are the larvae of winged fireflies. The Roman encyclopedist Pliny believed that fireflies turned their lights off and on by opening and closing their wings, a statement repeated again and again down through the Middle Ages, along with a great deal of other misinformation, including tales of luminous birds. Thomas Mouffet (1553–1604) was aware that the British glowworms were females and that the males were nonluminous flying insects. Like many persons of his time, Mouffet was most interested in the medical uses of plants and animals. Fireflies, says Mouffet, "being drank in wine make the use of lust not only irksome but loathsome. . . It were worthily wisht therefore that the unclean sort of Letchers were with the frequent taking of these in Potion disabled, who spare neither wife, widow nor maid, but defile themselves with lust not fit to be mentioned."

The scientific study of insects is sometimes said to have begun with the publication of Ulysses Aldrovandi's *De Animalibus Insectis* in Bologna in 1602. Aldrovandi included a fairly accurate sketch of a glowworm, as well as the interesting hypothesis that fireflies use their lights to find their way about at night. A few years later Francis Bacon expressed curiosity that these insects were able to produce light without heat, but the times were scarcely ripe for a solution to this problem. The first important book on animal lights was written by the Danish physician Thomas Bartholin in 1647. Bartholin's own experiments failed when his glowworms escaped from the cage, but he discussed the unpublished work of Vintimillia, an Italian who observed the mating of fireflies in glass jars. Vintimillia was well aware that the flashes serve to attract the sexes, and he was the first to note that the eggs are luminous. During the eighteenth and nineteenth centuries a great many persons turned their attention to the life histories and luminescent properties of fireflies, including such notables as Michael Faraday and Louis Pasteur. We nevertheless still have a long way to go; one can imagine a scientist of the year 2068 looking back to our time with somewhat the same amusement we now look back on Thomas Mouffet, although considering our increasing unmindfulness of the past, it is equally possible to imagine that in 2068 men will not look back at all. Or that there will not be men at all.

Perhaps the most notable contribution to an understanding of the light of fireflies was made in 1885 by the French physiologist Raphael Dubois. Dubois removed a light organ of the beetle Pyrophorus, ground it up in water, and left it until the light went out of its own accord. He then removed another organ and ground it in boiling water for a short time, so that its light, too, was extinguished. Then he performed a neat bit of magic: when the two extracts were placed together, the light reappeared. He deduced that two substances were required to produce light, one of which was inactivated by heat. He called these two substances luciferin and luciferase (after Lucifer, who among other more devilish traits was the bearer of light). Dubois also learned how to obtain luminous bacteria from the skins of dead fish and squids on the seashore. The bacteria could be transferred to culture plates, where they produced large colonies that glowed with a blue-green light. At the International Exposition in Paris in 1900, Dubois created a sensation by lighting a small room with flasks containing suspensions of these luminous bacteria.

A good deal more has now been learned about the production of animal light, and luciferin and luciferase have been obtained in purified crystalline form. McElroy and his colleagues at Johns Hopkins have synthesized luciferin. We now know that something more is needed: adenosine triphosphate (ATP). ATP may be less familiar to most persons than DDT or the CIA, but it happens to be even more important to us, providing as it does the energy for muscle contraction in animal bodies, including our own. In the light organ of the firefly, ATP energizes not muscles but the luciferin-luciferase system, the energy appearing not as mechanical work but as light. It has recently been proposed that luciferin and luciferase be employed in automated laboratories sent to Mars or other planets. The idea is that a scoop would pick up soil from the surface and mix it with water, oxygen, luciferin and luciferase. Then if a glow were televised back to earth, we would know that ATP, the fifth requirement for firefly-light production, occurs there. The presence of ATP would mean, in turn, the existence of some kind of animal life in that alien soil. Thoughts such as these emphasize the need for caution when labeling the study of fireflies (or anything else) "useless" or "idle curiosity."

In the living insect, an additional element is needed to account for the working of the system: some sort of nervous control. It was discovered long ago that cutting off the firefly's head caused the flashing to cease, although in some cases the light organ glows dimly for a long time. Later it was found that by electrical stimulation of the severed nerve cord one can produce experimental flashing. It is believed that nervous control is centered in the brain (much like the control of chirping in the cricket); impulses then travel to the light organs via the nerve cord and via delicate nerves that closely parallel the minute tubes that carry air to the light cells. We still do not know exactly how the flashing is triggered. Some have claimed that the nerves control the supply of oxygen to the light cells, but recent work suggests that the oxygen supply may be constant and that the series of chemical reactions resulting in a light flash may be initiated by the synaptic fluid of the nerve endings. That there are chemical intermediaries between nerve and light organ is suggested by the fact that a nervous shock provided directly to the light organ produces a very quick flash, whereas a stimulation to the nerves always involves a longer delay than nerve conduction itself would require. These are profound matters that we understand only poorly. Indeed, we still have much to learn as to how the chemical energy supplied by ATP is converted into the mechanical energy of ordinary muscle contraction.

The light organs of fireflies are complex structures, and recent studies using the electron microscope show them to be even more complex than once supposed. Each is composed of three layers: an outer "window," simply a transparent portion of the body wall; the light organ proper; and an inner layer of opaque, whitish cells filled with granules of uric acid, the so-called "reflector." The light organ proper contains large, slablike light cells, each of them filled with large granules and much smaller, dark granules, the latter tending to be concentrated around the numerous air tubes and nerves penetrating the light organ. These smaller granules were once assumed by some persons to be luminous bacteria, but we now know that they are mitochondria, the source of ATP and therefore of the energy of light production. The much larger granules that fill most of the light cells are still of unknown function; perhaps they serve as the source of luciferin.

Actually, the light organs vary a good deal in different kinds of fireflies. We also know that the color of the light varies in different species and that this is a real difference in light color and not the result of a tinting or filtering effect of the window. Generally speaking, the light is yellowish, but it may have a greenish, bluish, or orange hue. McElroy has found that the color of the light produced by luciferin can be changed by altering the alkalinity of the solution, less alkalinity producing a shift toward the red end of the spectrum. Present evidence suggests that various species of fireflies have slightly different luciferase molecules, which cause the production of light of slightly different wavelengths. In the genus Pyrophorus (not really a true firefly, but a click beetle) there are two greenish lights just behind the head and an orange light on the abdomen. I well remember my first acquaintance with Pyrophorus. We were camped out near the ruins of Xochicalco, in Morelos, Mexico, when a disturbance caused me to peer out into the darkness: only to find that we were surrounded by pairs of glowing green eyes. The ghosts of Toltec warriors a few yards away? No, it proved to be a host of Pyrophorus in the bushes only a foot or two away. The story is told that when Sir Robert Dudley and Sir James Cavendish first landed in Cuba, they saw great numbers of lights moving about in the woods. Supposing them to be Spaniards with torches, ready to advance upon them, the British withdrew to their ships and went on to settle Jamaica. In this manner Pyrophorus may be said to have changed the course of history.

The South American "railroad worm" is an elongate glowworm

having eleven greenish lights down each side of the body and two red lights on the head. These lights are quite brilliant, and when the insect is moving along the ground it looks like nothing so much as a fully lighted railroad train. The North American railroad worm is larger but lacks the red lights on the head. Both of these insects are quite rare.

We now know that there are not only differences in the nature, shape, and position of the light organs and in the color of the light of fireflies but also (and most particularly) in the behavior patterns of the male and female during courtship and mating. The males of the European glowworm fly toward a light only if it is of the shape, color, and intensity of that of the female of that species. In our common North American species, the females often rest on the ground or vegetation, and flash only in response to the flashes of the males. In one of the best-studied forms, Photinus pyralis, the male flies near the ground in a strongly undulating pattern; he approaches the bottom of one of these undulations every six seconds, and as he does so he makes a half-second flash, at the same time rising and thus describing a "J" of yellow-green light. If he passes within a few feet of a female, the latter responds with a half-second flash of her own, but only after an interval of about two seconds (with only slight variation). This interval is an all-important signal to the male; we know this because the male will respond to various flashes, including even that of a flashlight, but *only* when these occur about two seconds after his own flash. If the female flashes at the proper interval, he flies toward her and flashes again, whereupon the female again responds in two seconds. This may be repeated several times until the male reaches the female and mates with her. There is no evidence that sound or smell play any role in firefly mating.

The larger fireflies of eastern North America belong mostly to the genus Photuris, a confusing group in which the males show much variation in flash pattern but hardly any differences in structure or body color. For many years this problem bothered H. S. Barber, beetle specialist of the United States National Museum (not to be confused with H. G. Barber, a specialist on true bugs who worked at the National Museum at the same time—the two were "beetle Barber" and "bug Barber" to their colleagues). The results of H. S. Barber's study were not published until a year

after his death in 1950. Barber found that in the Potomac Valley
he could detect a woodland species with a short greenish-white
flash once a second; a stream-side species with a slightly slower,
faintly orange flash; a species occurring in alder groves and poising almost motionless, its light beginning dimly and growing
steadily in brilliance before stopping abruptly, only to reappear
at a different point several seconds later; and so forth. Eventually
Barber recognized eighteen species of Photuris, mainly on the basis
of the flashes of the males; ten of these he had to name as new,
since they had not previously been recognized. Needless to say,
this did not endear him to museum workers, who could not very
well sort their dead beetles on the basis of their flashes. But as
Barber said:

> Taxonomy from old mummies which fill collections is a
> misguided concept. It leads to the misidentification of rot-
> ten old samples in collections. How these poor fireflies
> would resent being placed in such diverse company—among
> specimens of enemy species—if they were alive and intelli-
> gent! What contempt they would feel for the "damned
> taxonomist."

Dr. James E. Lloyd, of the University of Florida, has recently
completed a study of flash communication in the genus Photinus,
the common smaller fireflies of the eastern United States. (Did you
know that the Pacific coastal states, despite their many attractions,
have almost no fireflies?) Lloyd, too, found several "hidden" species, first recognized by consistent differences in flash signals, and
later found to differ in minor details of body color. In many places
two or more species of Photinus fly together, but they are prevented from interbreeding by their different light signals. The
males fly at different heights and in different flight patterns; their
flashes differ in length, in the number of pulses per flash, and
sometimes in the color or intensity of the light. The male is saying,
in Lloyd's words: "Here I am in time and space, a sexually mature
male of species X that is ready to mate. Over." The female of "species X" responds with a flash at the interval characteristic of her
species—as described above for Photinus pyralis. Lloyd was interested in learning how much latitude was permissible without
causing a "misunderstanding." In his experiments he used elec-
tronic devices for producing artificial flashes of known duration

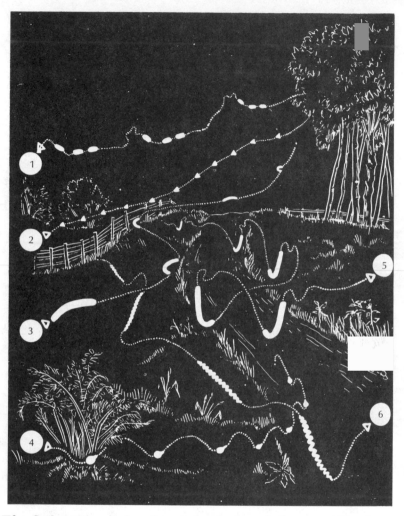

The flash patterns of males of six species of the genus Photinus. Each is signaling to the female of his own species in his own characteristic pattern. No. 1 flies two to four yards high and produces three slow flashes in series. No. 2 flies in a rather straight path somewhat lower, producing single flashes that increase in intensity during emission. No. 3 is a low flier that emits a long flash while executing a lateral curve, while no. 4 makes a series of hops between which he hovers and produces a quick flash. No. 5 is as well as for accurately measuring the response delay of the females. As in the case of crickets—and in fact all "cold-blooded" animals—things happen faster at higher temperatures, so in all his work temperature had to be taken into account.

In any given locality, the males and females are highly attuned to one another's messages; that is, the variation in responsiveness is such that they almost never answer another species. Females occasionally reply once to a flash of inappropriate length, but they do not continue to do so. On the other hand, if one compares the flash signals of species that do not occur together he often finds them to be very similar: here there is no possibility of mistakes being made, and refined "isolating mechanisms" have not evolved. It goes without saying that the integrity of species must be maintained, for interspecies hybrids are generally sterile (like the mule) or at least less well adapted for a specific role in nature.

One would assume that the larger fireflies of the genus Photuris (studied by Barber) always "speak a different language" from the small fireflies of the genus Photinus (studied by Lloyd). This is, of course, generally so, but with some fascinating exceptions. H. S. Barber commented on this as follows:

> Sometimes the familiar flashes of a small species of Photinus male are observed excitedly courting a female, supposedly of the same species, whose flashes appear normal to its kind, but when the electric light is thrown upon them one is startled to find the intended bride of the Photinus is a large and very alert female Photuris facing him with great interest. Does she lure him to serve as her repast? Very often a dim steady light near the ground proves under the flashlamp to be a small, recently killed Photinus being devoured by a nonluminous female Photuris. . . .

James Lloyd, while working on Photinus, found it possible to obtain females of a given species by walking about in a suitable habitat, imitating the flashes of the males with a flashlight. But

---

Photinus pyralis with its characteristic J-shaped signal, as discussed in the text. No. 6 is a low-flying species which produces a long flash while jerking rapidly from side to side. (Adapted from a drawing by Daniel Otte in James Lloyd's study of Photinus fireflies.)

now and then the females that signaled back to him turned out
not to be Photinus females, but those of the genus Photuris, re-
sponding appropriately to specific signals of a certain species of
Photinus! Once he watched one of the Photuris females for half
an hour and saw her respond to twelve passing Photinus males, in
each case after the interval characteristic of that species of Photinus; all of these males were at least partially attracted to her. Finally a male landed near her, and after an exchange of signals
ceased to light up after the usual time period. Lloyd checked and
found that the Photuris female was clasping the Photinus male
and chewing on him. As Lloyd points out:

> The answer to Barber's question has precipitated a deluge
> of new questions, not the least of which concerns the males
> of the genus Photuris. Is the female Photuris predaceous be-
> fore she has mated? If so, how does her mate avoid the fate
> of attracted Photinus males? . . . Can a single Photuris
> species prey upon more than one Photinus species with dif-
> ferent signal systems? In other words, how many flash pat-
> terns do Photuris females have in their "repertories," and is
> predation on Photinus fireflies in any sense obligatory?

It might be added parenthetically that insects are known that
utilize luminescence not for courtship but strictly for luring and
then feeding upon various small insects that are naturally attracted to light. Both in North America and in Europe there are
certain gnat larvae that spin silken webs close to the ground and
emit a dim, bluish light that probably serves to attract tiny midges
and other insects into the web. An even better example is provided by the so-called New Zealand glowworm, which is not a true
glowworm at all but another gnat larva. These insects live in certain caves in New Zealand and are so spectacular that guided tours
are conducted into some of the caves. The gnats lay their eggs in a
gluelike substance on the ceiling, and the larvae suspend themselves from silken sheaths and emit a bluish-green light that is said
to lure small insects into the tangle of webs, where they are con-
sumed by the larvae. F. W. Edwards describes the experience of
entering the depths of one of these caves as follows:

> [After being warned by the guide to be quiet] we stepped
> cautiously in single file down, down to a still lower level.

> . . . Then gradually we became aware that a vision was
> silently breaking on us . . . a radiance became manifest
> which absorbed the whole faculty of observations—the radi-
> ance of such a massed body of glowworms as cannot be
> found anywhere else in the world, utterly incalculable as to
> numbers and merging their individual lights in a nirvana
> of pure sheen.

True fireflies are also capable of remarkable displays at times.
Occasionally (especially in the tropics) untold thousands of fireflies
will gather in a single tree or several neighboring trees and flash
for many hours, sometimes for many nights in succession, producing a glow that can be seen half a mile or more away. Sometimes
all fireflies in a tree have been seen to flash in synchrony. Such
displays have been reported from Southeast Asia and the East Indies for over two hundred years—but hardly ever from other parts
of the world. Hugh M. Smith, while studying the fisheries of Thailand in the 1930's, often took parties of visitors down the Chao
Phraya River near Bangkok to observe the displays. In an article
in *Science,* he described them in these words:

> Imagine a tree thirty-five to forty feet high thickly covered
> with small ovate leaves, apparently with a firefly on every
> leaf and all the fireflies flashing in perfect unison at the rate
> of about three times in two seconds, the tree being in com-
> plete darkness between the flashes. . . . Imagine a tenth of a
> mile of river front with an unbroken line of [mangrove]
> trees with fireflies on every leaf flashing in unison. . . .
> Then, if one's imagination is sufficiently vivid, he may form
> some conception of this amazing spectacle.

Smith went on to say that the synchronous flashing occurs "hour
after hour, night after night, for weeks or even months. . . ." Reports such as Smith's have tended to remove much of the skepticism
that greeted earlier accounts. (An author of an article in *Science*
some years earlier had attributed the flashing to the twitching of
the eyelids, remarking that "the insects had nothing whatever to
do with it"!) For years the explanation of this unique phenomenon has intrigued John Buck, of the National Institutes of Health
at Bethesda, Maryland, one of our leading authorities on fireflies.

Some time ago he found that he could induce synchronous flashing on a small scale in the American firefly Photinus pyralis by using a flashlight at the usual interval of females of this species. When there were many males about, he could sometimes attract fifteen or twenty of them at once, and these would all adjust their flash periodicity in accordance with that of the female. "It is indeed an impressive sight," says Buck, "to see such a group converging through the air toward one point, each member poising, flashing and surging forward in short advances, all in the most perfect synchronism." It seemed possible that small groups such as this might build up within a larger aggregation and so stimulate one another that all fell into synchrony.

In another experiment, Dr. Buck placed a large number of males of this same species in a large, dimly lighted cage, where they soon began to flash in their usual manner. He then subjected the fireflies to sudden and complete darkness, whereupon all of them flashed at once, then again after four or five seconds. The synchrony persisted for some time and then disappeared. Buck felt that the unnatural advent of sudden, total darkness was not of importance in itself, but only because it served to increase the relative intensity of the flashes of neighboring fireflies, causing them to respond to one another's flashes as they would not ordinarily do in nature.

But of course these simple experiments performed on a North American species merely whetted his appetite for the real thing, and a couple of years ago John and Elisabeth Buck took off for Thailand and Borneo. They were successful in finding "firefly trees," and they made photographic and photometric analyses that indicate that synchrony of great numbers of individuals is indeed nearly perfect. They found that (contrary to earlier reports) both males and females occur in these trees, although the females do not participate in synchronous flashing. They showed that mating occurs in the trees, suggesting that the brilliant, synchronous flashes serve as a beacon to attract females from the surrounding forest. This may explain why this phenomenon is most prevalent along rivers in the Far East, for in this part of the world the exceedingly dense, tangled swamps would hardly be conducive to individual flash communication similar to that occurring in a New England meadow. But a "firefly tree" along a watercourse would provide an assembly beacon of ready access. Not only would the synchrony of the flashes increase the brightness but the alternation of light and dark would also be eye-catching, like the flashing neon signs that are a recent invention of man (though I would think that man has overdone a good thing as he so often does).

The Bucks consider synchronous flashing to be a complex of behavior patterns (congregation, selection of certain trees, flashing, synchrony, and so forth) that have evolved together into a spectacular device for enhancing mating under otherwise difficult conditions. Evidently newly emerged males and females are constantly recruited from the surrounding forests, for individuals do not live more than a few days, and there must be a constant turnover in the population. The Bucks showed that males released in a darkroom are attracted to each other's light, and this suggests that wandering individuals might readily join a flashing swarm. It remains to be proved that there is a traffic of freshly emerged males and females into the trees and of mated females away from them. And it remains to be shown how the males maintain almost perfect synchrony from one end of a large aggregation to the other, when in fact laboratory studies suggest that the males react to one another over only short distances and that their reaction time is considerably greater than the variation in synchrony observed in nature. There is evidence that near-perfect synchrony occurs only in very dense aggregations, while in diffuse gatherings the flashing may be random. In some instances more than one species may aggregate in a given tree, resulting in a complex combination of flashes that is still presumably effective in attracting females of the species involved. All these are matters requiring much further study.

But of course scientists are used to partial and provisional answers; it is their stock in trade, and half the fun of science. H. S. Barber was well aware that his field studies of Photuris were only preliminary. And after a lengthy review of laboratory studies, John Buck concluded:

> In spite of the many morphological and physiological data which concern luminescence in the firefly, there seem to be surprisingly few unequivocal major conclusions which can be drawn.

This is "par for the course." Such is the complexity of living systems that tens of thousands of research workers all over the world each year push our knowledge forward by only a minuscule, with now and then a breakhrough that opens up a new area of ignorance. A century from now our great-grandchildren may marvel at how little we knew about fireflies. At least I hope so. In the meantime we may be unashamedly romantic or unflinchingly rigorous in our attitude toward fireflies, as befits our nature, and still know their magic.

QUESTIONS AND SUGGESTIONS

1. Evans is really writing an informal review article. That is, he is surveying the contributions of several scientists to one problem, that of luminescence in insects, especially in Photinus, the firefly. Note how he incorporates the work of others into his own framework of exposition, especially through his use of the first person. Point out some of the stylistic features that contribute to the informality of his tone.

2. In the course of the article, he communicates much more than facts; we come to know a great deal about him, about his sense of nature. Describe Evans's attitude toward his subject; toward science.

3. Does science make the mysterious less mysterious to you?

4. Write an informal review article on one problem with which you are familiar. To what extent does your choice of subject determine your audience? Why are fireflies a fit subject for a *New Yorker* article?

LEWIS THOMAS

Vibes

LEWIS THOMAS (b. 1913) is a research physician who has specialized in infectious diseases, hypersensitivity, and the pathogenicity of mycoplasmas. As President of the Memorial Sloan-Kettering Cancer Center, he continues a distinguished career in research and teaching. Originally appearing in the *New England Journal of Medicine,* the essays later collected as *Lives of a Cell: Notes of a Biology-Watcher* had great popular appeal and won the National Book Award in 1975; two of the essays are reprinted here. Lewis Thomas is often compared to Sir Thomas Browne, a seventeenth-century prose stylist, also a physician, who wrote on natural philosophy (the century's term for "science") and made it the subject of spiritual meditation.

We leave traces of ourselves wherever we go, on whatever we touch. One of the odd discoveries made by small boys is that when two pebbles are struck sharply against each other they emit, briefly, a curious smoky odor. The phenomenon fades when the stones are immaculately cleaned, vanishes when they are heated to furnace temperature, and reappears when they are simply touched by the hand again before being struck.

An intelligent dog with a good nose can track a man across open ground by his smell and distinguish that man's tracks from those of others. More than this, the dog can detect the odor of a light human fingerprint on a glass slide, and he will remember that slide and smell it out from others for as long as six weeks, when the scent fades away. Moreover, this animal can smell the identity of identical twins, and will follow the tracks of one or the other as though they had been made by the same man.

We are marked as self by the chemicals we leave beneath the soles of our shoes, as unmistakably and individually as by the membrane surface antigens detectable in homografts of our tissues.

Other animals are similarly endowed with signaling mechanisms. Columns of ants can smell out the differences between themselves and other ants on their trails. The ants of one species, proceeding jerkily across a path, leave trails that can be followed by their own relatives but not by others. Certain ants, predators, have taken unfair advantage of the system; they are born with an ability to sense the trails of the species they habitually take for slaves, follow their victims to their nests, and release special odorants that throw them into disorganized panic.

Minnows and catfish can recognize each member of their own species by his particular, person-specific odor. It is hard to imagine a solitary, independent, existentialist minnow, recognizable for himself alone; minnows in a school behave like interchangeable, identical parts of an organism. But there it is.

The problem of olfactory sensing shares some of the current puzzles and confusions of immunology, apart from the business of telling self from non-self. A rabbit, it has been calculated, has something like 100 million olfactory receptors. There is a constant and surprisingly rapid turnover of the receptor cells, with new ones emerging from basal cells within a few days. The theories to explain olfaction are as numerous and complex as those for immunologic sensing. It seems likely that the shape of the smelled molecule is what matters most. By and large, odorants are chemically small, Spartan compounds. In a rose garden, a rose is a rose because of geraniol, a 10-carbon compound, and it is the geometric conformation of atoms and their bond angles that determine the unique fragrance. The special vibrations of atoms or groups of atoms within the molecules of odorants, or the vibratory song of the entire molecule, have been made the basis for several theories, with postulated "osmic frequencies" as the source of odor. The geometry of the molecule seems to be more important than the names of the atoms themselves; any set of atoms, if arranged in precisely the same configuration, by whatever chemical name, might smell as sweet. It is not known how the olfactory cells are fired by an odorant. According to one view, a hole is poked in the receptor membrane, launching depolarization, but other workers believe that the substance may become bound to the cells possessing specific receptors for it and then may just sit there, somehow displaying its signal from a distance, after the fashion of antigens on immune cells. Specific receptor proteins have been proposed, with different olfactory cells carrying specific receptors for different "primary" odors, but no one has yet succeeded in identifying the receptors or naming the "primary" odors.

Training of cells for olfactory sensing appears to be an every-day phenomenon. Repeated exposure of an animal to the same odorant, in small doses, leads to great enhancement of acuity, suggesting the possibility that new receptor sites are added to the cells. It is conceivable that new clones of cells with a particular receptor are stimulated to emerge in the process of training. The guinea pig, that immunologically famous animal, can be trained to perceive fantastically small amounts of nitrobenzene by his nose, without the help of Freund's adjuvants or haptene carriers. Minnows have been trained to recognize phenol, and distinguish it from p-chlorophenol, in concentrations of five parts per billion. Eels have been taught to smell two or three molecules of phenyl-ethyl alcohol. And, of course, eels and salmon must be able to re-member by nature, as the phrase goes, the odor of the waters in which they were hatched, so as to sniff their way back from the open sea for spawning. Electrodes in the olfactory bulbs of salmon will fire when the olfactory epithelium is exposed to water from their spawning grounds, whereas water from other streams causes no response.

We feel somehow inferior and left out of things by all the mar-velous sensory technology in the creatures around us. We some-times try to diminish our sense of loss (or loss of sense) by claiming to ourselves that we have put such primitive mechanisms behind us in our evolution. We like to regard the olfactory bulb as a sort of archeologic find, and we speak of the ancient olfactory parts of the brain as though they were elderly, dotty relatives in need of hobbies.

But we may be better at it than we think. An average man can detect just a few molecules of butyl mercaptan, and most of us can sense the presence of musk in vanishingly small amounts. Steroids are marvelously odorous, emitting varieties of musky, sexy smells. Women are acutely aware of the odor of a synthetic steroid named exaltolide, which most men are unable to detect. All of us are able to smell ants, for which the great word pismire was originally coined.

There may even be odorants that fire off receptors in our olfac-tory epithelia without our being conscious of smell, including sig-nals exchanged involuntarily between human beings. Wiener has proposed, on intuitive grounds, that defects and misinterpreta-

tions in such a communication system may be an unexplored territory for psychiatry. The schizophrenic, he suggests, may have his
problems with identity and reality because of flawed perceptions
of his own or others' signals. And, indeed, there may be something
wrong with the apparatus in schizophrenics; they have, it is said,
an unfamiliar odor, recently attributed to trans-3-methyl-hexanoic
acid, in their sweat.

Olfactory receptors for communication between different creatures are crucial for the establishment of symbiotic relations. The
crab and anemone recognize each other as partners by molecular
configurations, as do the anemones and their symbiotic damsel
fish. Similar devices are employed for defense, as with the limpet,
which defends itself against starfish predators by everting its man-
tle and thus precluding a starfish foothold; the limpet senses a
special starfish protein, which is, perhaps in the name of fairness,
elaborated by all starfish into their environment. The system is
evidently an ancient one, long antedating the immunologic sensing of familiar or foreign forms of life by the antibodies on which
we now depend so heavily for our separateness. It has recently
been learned that the genes for the marking of self by cellular antigens and those for making immunologic responses by antibody
formation are closely linked. It is possible that the invention of
antibodies evolved from the earlier sensing mechanisms needed
for symbiosis, perhaps designed, in part, to keep the latter from
getting out of hand.

A very general system of chemical communication between living things of all kinds, plant and animal, has been termed "allelo-
chemics" by Whittaker. Using one signal or another, each form of
life announces its proximity to the others around it, setting limits
on encroachment or spreading welcome to potential symbionts.
The net effect is a coordinated mechanism for the regulation of
rates of growth and occupations of territory. It is evidently designed for the homeostasis of the earth.

Jorge Borges, in his recent bestiary of mythical creatures, notes
that the idea of round beasts was imagined by many speculative
minds, and Johannes Kepler once argued that the earth itself is
such a being. In this immense organism, chemical signals might
serve the function of global hormones, keeping balance and symmetry in the operation of various interrelated working parts, in-

forming tissues in the vegetation of the Alps about the state of eels in the Sargasso Sea, by long, interminable relays of interconnected messages between all kinds of other creatures.

This is an interesting kind of problem, made to order for computers if they came in sizes big enough to store in nearby galaxies. It is nice to think that there are so many unsolved puzzles ahead for biology, although I wonder whether we will ever find enough graduate students.

[For questions and suggestions, see pages 138–39.]

LEWIS THOMAS

Germs

Watching television, you'd think we lived at bay, in total jeopardy, surrounded on all sides by human-seeking germs, shielded against infection and death only by a chemical technology that enables us to keep killing them off. We are instructed to spray disinfectants everywhere, into the air of our bedrooms and kitchens and with special energy into bathrooms, since it is our very own germs that seem the worst kind. We explode clouds of aerosol, mixed for good luck with deodorants, into our noses, mouths, underarms, privileged crannies—even into the intimate insides of our telephones. We apply potent antibiotics to minor scratches and seal them with plastic. Plastic is the new protector; we wrap the already plastic tumblers of hotels in more plastic, and seal the toilet seats like state secrets after irradiating them with ultraviolet light. We live in a world where the microbes are always trying to get at us, to tear us cell from cell, and we only stay alive and whole through diligence and fear.

We still think of human disease as the work of an organized, modernized kind of demonology, in which the bacteria are the most visible and centrally placed of our adversaries. We assume that they must somehow relish what they do. They come after us for profit, and there are so many of them that disease seems inevitable, a natural part of the human condition; if we succeed in eliminating one kind of disease there will always be a new one at hand, waiting to take its place.

These are paranoid delusions on a societal scale, explainable in part by our need for enemies, and in part by our memory of what things used to be like. Until a few decades ago, bacteria were a genuine household threat, and although most of us survived them, we were always aware of the nearness of death. We moved, with our families, in and out of death. We had lobar pneumonia, meningococcal meningitis, streptococcal infections, diphtheria, endocarditis, enteric fevers, various septicemias, syphilis, and, always, everywhere, tuberculosis. Most of these have now left most of us, thanks to antibiotics, plumbing, civilization, and money, but we remember.

In real life, however, even in our worst circumstances we have always been a relatively minor interest of the vast microbial world. Pathogenicity is not the rule. Indeed, it occurs so infrequently and involves such a relatively small number of species, considering the huge population of bacteria on the earth, that it has a freakish aspect. Disease usually results from inconclusive negotiations for symbiosis, an overstepping of the line by one side or the other, a biologic misinterpretation of borders.

Some bacteria are only harmful to us when they make exotoxins, and they only do this when they are, in a sense, diseased themselves. The toxins of diphtheria bacilli and streptococci are produced when the organisms have been infected by bacteriophage; it is the virus that provides the code for toxin. Uninfected bacteria are uninformed. When we catch diphtheria it is a virus infection, but not of us. Our involvement is not that of an adversary in a straightforward game, but more like blundering into someone else's accident.

I can think of a few microorganisms, possibly the tubercle bacillus, the syphilis spirochete, the malarial parasite, and a few others, that have a selective advantage in their ability to infect human beings, but there is nothing to be gained, in an evolutionary sense, by the capacity to cause illness or death. Pathogenicity may be something of a disadvantage for most microbes, carrying lethal risks more frightening to them than to us. The man who catches a meningococcus is in considerably less danger for his life, even without chemotherapy, than meningococci with the bad luck

JAMES WATSON

Finding the Secret of Life

JAMES WATSON (b. 1928) and Francis Crick (b. 1916) shared the 1962 Nobel Prize for physiology or medicine with M. H. Wilkins. Watson, an American, and Crick, an Englishman, began collaboration on the DNA structure in 1951 at Cambridge, and, after an unsuccessful first effort, they succeeded, in 1953, in devising a structure based on a complementary double helical configuration, which they proposed in an article in *Nature*. (The article is reprinted in this anthology; see pages 356–360.) Their work benefited greatly from the X-ray diffraction pictures made by Rosalind Franklin. Not only did they win a Nobel Prize, but Watson's account of the race to unlock the DNA code has been acclaimed as an accurate description of what doing science is "really like." Reprinted here are chapters from that account, *The Double Helix*, which describe the events leading to their representation of the DNA structure. *The Double Helix* was written to be a best-seller, and it succeeded.

Bragg was in Max's office when I rushed in the next day to blurt out what I had learned. Francis was not yet in, for it was a Saturday morning and he was still home in bed glancing at the *Nature* that had come in the morning mail. Quickly I started to run through the details of the B form, making a rough sketch to show the evidence that DNA was a helix which repeated its pattern every 34 Å along the helical axis. Bragg soon interrupted me with a question, and I knew my argument had got across. I thus wasted no time in bringing up the problem of Linus, giving the opinion that he was far too dangerous to be allowed a second crack at DNA while the people on this side of the Atlantic sat on their hands. After saying that I was going to ask a Cavendish machinist to make models of the purines and pyrimidines, I remained silent, waiting for Bragg's thoughts to congeal.

To my relief, Sir Lawrence not only made no objection but encouraged me to get on with the job of building models. He clearly was not in sympathy with the internal squabbling at King's—especially when it might allow Linus, of all people, to get the thrill of discovering the structure of still another important molecule. Also aiding our cause was my work on tobacco mosaic virus. It had

140

given Bragg the impression that I was on my own. Thus he could fall asleep that night untroubled by the nightmare that he had given Crick carte blanche for another foray into frenzied inconsiderateness. I then dashed down the stairs to the machine shop to warn them that I was about to draw up plans for models wanted within a week.

Shortly after I was back in our office, Francis strolled in to report that their last night's dinner party was a smashing success. Odile was positively enchanted with the French boy that my sister had brought along. A month previously Elizabeth had arrived for an indefinite stay on her way back to the States. Luckily I could both install her in Camille Prior's boarding house and arrange to take my evening meals there with Pop and her foreign girls. Thus in one blow Elizabeth had been saved from typical English digs, while I looked forward to a lessening of my stomach pains.

Also living at Pop's was Bertrand Fourcade, the most beautiful male, if not person, in Cambridge. Bertrand, then visiting for a few months to perfect his English, was not unconscious of his unusual beauty and so welcomed the companionship of a girl whose dress was not in shocking contrast with his well-cut clothes. As soon as I had mentioned that we knew the handsome foreigner, Odile expressed delight. She, like many Cambridge women, could not take her eyes off Bertrand whenever she spotted him walking down King's Parade or standing about looking very well-favored during the intermissions of plays at the amateur dramatic club. Elizabeth was thus given the task of seeing whether Bertrand would be free to join us for a meal with the Cricks at Portugal Place. The time finally arranged, however, had overlapped my visit to London. When I was watching Maurice meticulously finish all the food on his plate, Odile was admiring Bertrand's perfectly proportioned face as he spoke of his problems choosing among potential social engagements during his forthcoming summer on the Riviera.

This morning Francis saw that I did not have my usual interest in the French moneyed gentry. Instead, for a moment he feared that I was going to be unusually tiresome. Reporting that even a former birdwatcher could now solve DNA was not the way to greet a friend bearing a slight hangover. However, as soon as I revealed the B-pattern details, he knew I was not pulling his leg. Especially important was my insistence that the meridional reflection at 3.4 Å was much stronger than any other reflection. This could only mean that the 3.4 Å-thick purine and pyrimidine bases were stacked on top of each other in a direction perpendicular to the helical axis. In addition we could feel sure from both electron-microscope and X-ray evidence that the helix diameter was about 20 Å.

Francis, however, drew the line against accepting my assertion that the repeated finding of twoness in biological systems told us to build two-chain models. The way to get on, in his opinion, was to reject any argument which did not arise from the chemistry of nucleic-acid chains. Since the experimental evidence known to us could not yet distinguish between two- and three-chain models, he wanted to pay equal attention to both alternatives. Though I remained totally skeptical, I saw no reason to contest his words. I would of course start playing with two-chain models.

No serious models were built, however, for several days. Not only did we lack the purine and pyrimidine components, but we had never had the shop put together any phosphorus atoms. Since our machinist needed at least three days merely to turn out the more simple phosphorus atoms, I went back to Clare after lunch to hammer out the final draft of my genetics manuscript. Later, when I cycled over to Pop's for dinner, I found Bertrand and my sister talking to Peter Pauling, who the week before had charmed Pop into giving him dining rights. In contrast to Peter, who was complaining that the Perutzes had no right to keep Nina home on a Saturday night, Bertrand and Elizabeth looked pleased with themselves. They had just returned from motoring in a friend's Rolls to a celebrated country house near Bedford. Their host, an anti-quarian architect, had never truckled under to modern civilization and kept his house free of gas and electricity. In all ways possible he maintained the life of an eighteenth-century squire, even to providing special walking sticks for his guests as they accompanied him around his grounds.

Dinner was hardly over before Bertrand whisked Elizabeth on to another party, leaving Peter and me at a loss for something to do. After first deciding to work on his hi-fi set, Peter came along with me to a film. This kept us in check until, as midnight approached, Peter held forth on how Lord Rothschild was avoiding his responsibility as a father by not inviting him to dinner with his daughter Sarah. I could not disagree, for if Peter moved into the fashionable world I might have a chance to escape acquiring a faculty-type wife.

Three days later the phosphorus atoms were ready, and I quickly strung together several short sections of the sugar-phosphate backbone. Then for a day and a half I tried to find a suitable two-chain model with the backbone in the center. All the possible models compatible with the B-form X-ray data, however, looked stereochemically even more unsatisfactory than our three-chained models of fifteen months before. So, seeing Francis absorbed by his thesis, I took off the afternoon to play tennis with Bertrand. After tea I returned to point out that it was lucky I found tennis more pleasing than model building. Francis, totally indifferent to the perfect spring day, immediately put down his pencil to point out that not only was DNA very important, but he could assure me that someday I would discover the unsatisfactory nature of outdoor games.

During dinner at Portugal Place I was back in a mood to worry about what was wrong. Though I kept insisting that we should keep the backbone in the center, I knew none of my reasons held water. Finally over coffee I admitted that my reluctance to place the bases inside partially arose from the suspicion that it would be possible to build an almost infinite number of models of this type. Then we would have the impossible task of deciding whether one was right. But the real stumbling block was the bases. As long as they were outside, we did not have to consider them. If they were pushed inside, the frightful problem existed of how to pack together two or more chains with irregular sequences of bases. Here Francis had to admit that he saw not the slightest ray of light. So when I walked up out of their basement dining room into the street, I left Francis with the impression that he would have to provide at least a semiplausible argument before I would seriously play about with base-centered models.

The next morning, however, as I took apart a particularly repulsive backbone-centered molecule, I decided that no harm could come from spending a few days building backbone-out models. This meant temporarily ignoring the bases, but in any case this had to happen since now another week was required before the shop could hand over the flat tin plates cut in the shapes of purines and pyrimidines.

There was no difficulty in twisting an externally situated backbone into a shape compatible with the X-ray evidence. In fact, both Francis and I had the impression that the most satisfactory angle of rotation between two adjacent bases was between 30 and 40 degrees. In contrast, an angle either twice as large or twice as small looked incompatible with the relevant bond angles. So if the backbone was on the outside, the crystallographic repeat of 34 Å had to represent the distance along the helical axis required for a complete rotation. At this stage Francis' interest began to perk up, and at increasing frequencies he would look up from his calculations to glance at the model. Nonetheless, neither of us had any hesitation in breaking off work for the weekend. There was a party at Trinity on Saturday night, and on Sunday Maurice was coming up to the Cricks' for a social visit arranged weeks before the arrival of the Pauling manuscript.

Maurice, however, was not allowed to forget DNA. Almost as soon as he arrived from the station, Francis started to probe him for fuller details of the B pattern. But by the end of lunch Francis knew no more than I had picked up the week before. Even the presence of Peter, saying he felt sure his father would soon spring into action, failed to ruffle Maurice's plans. Again he emphasized that he wanted to put off more model building until Rosy was gone, six weeks from then. Francis seized the occasion to ask Maurice whether he would mind if we started to play about with DNA models. When Maurice's slow answer emerged as no, he wouldn't mind, my pulse rate returned to normal. For even if the answer had been yes, our model building would have gone ahead.

The next few days saw Francis becoming increasingly agitated by my failure to stick close to the molecular models. It did not matter that before his tenish entrance I was usually in the lab. Almost every afternoon, knowing that I was on the tennis court, he would fretfully twist his head away from his work to see the polynucleotide backbone unattended. Moreover, after tea I would show up for only a few minutes of minor fiddling before dashing away to have sherry with the girls at Pop's. Francis' grumbles did not disturb me, however, because further refining of our latest backbone without a solution to the bases would not represent a real step forward.

I went ahead spending most evenings at the films, vaguely dreaming that any moment the answer would suddenly hit me. Occasionally my wild pursuit of the celluloid backfired, the worst occasion being an evening set aside for *Ecstasy*. Peter and I had both been too young to observe the original showings of Hedy Lamarr's romps in the nude, and so on the long-awaited night we collected Elizabeth and went up to the Rex. However, the only swimming scene left intact by the English censor was an inverted reflection from a pool of water. Before the film was half over we joined the violent booing of the disgusted undergraduates as the dubbed voices uttered words of uncontrolled passion.

Even during good films I found it almost impossible to forget the bases. The fact that we had at last produced a stereochemically reasonable configuration for the backbone was always in the back of my head. Moreover, there was no longer any fear that it would be incompatible with the experimental data. By then it had been checked out with Rosy's precise measurements. Rosy, of course, did not directly give us her data. For that matter, no one at King's realized they were in our hands. We came upon them because of Max's membership on a committee appointed by the Medical Research Council to look into the research activities of Randall's lab to coordinate Biophysics research within its laboratories. Since Randall wished to convince the outside committee that he had a productive research group, he had instructed his people to draw up a comprehensive summary of their accomplishments. In due time this was prepared in mimeograph form and sent routinely to all the committee members. The report was not confidential and so Max saw no reason not to give it to Francis and me. Quickly scanning its contents, Francis sensed with relief that following my return from King's I had correctly reported to him the essential features of the B pattern. Thus only minor modifications were necessary in our backbone configuration.

Generally, it was late in the evening after I got back to my rooms that I tried to puzzle out the mystery of the bases. Their formulas were written out in J. N. Davidson's little book *The Biochemistry of Nucleic Acids,* a copy of which I kept in Clare. So

I could be sure that I had the correct structures when I drew tiny pictures of the bases on sheets of Cavendish notepaper. My aim was somehow to arrange the centrally located bases in such a way that the backbones on the outside were completely regular—that is, giving the sugar-phosphate groups of each nucleotide identical three-dimensional configurations. But each time I tried to come up with a solution I ran into the obstacle that the four bases each had a quite different shape. Moreover, there were many reasons to believe that the sequences of the bases of a given polynucleotide chain were very irregular. Thus, unless some very special trick existed, randomly twisting two ploynucleotide chains around one another should result in a mess. In some places the bigger bases must touch each other, while in other regions, where the smaller bases would lie opposite each other, there must exist a gap or else their backbone regions must buckle in.

There was also the vexing problem of how the intertwined chains might be held together by hydrogen bonds between the bases. Though for over a year Francis and I had dismissed the possibility that bases formed regular hydrogen bonds, it was now obvious to me that we had done so incorrectly. The observation that one or more hydrogen atoms on each of the bases could move from one location to another (a tautomeric shift) had initially led us to conclude that all the possible tautomeric forms of a given base occurred in equal frequencies. But a recent rereading of J. M. Gulland's and D. O. Jordan's papers on the acid and base titrations of DNA made me finally appreciate the strength of their conclusion that a large fraction, if not all, of the bases formed hydrogen bonds to other bases. Even more important, these hydrogen bonds were present at very low DNA concentrations, strongly hinting that the bonds linked together bases in the same molecule. There was in addition the X-ray crystallographic result that each pure base so far examined formed as many irregular hydrogen bonds as stereochemically possible. Thus, conceivably the crux of the matter was a rule governing hydrogen bonding between bases.

My doodling of the bases on paper at first got nowhere, regardless of whether or not I had been to a film. Even the necessity to expunge *Ecstasy* from my mind did not lead to passable hydrogen bonds, and I fell asleep hoping that an undergraduate party the next afternoon at Downing would be full of pretty girls. But my expectations were dashed as soon as I arrived to spot a group of healthy hockey players and several pallid debutantes. Bertrand also instantly perceived he was out of place, and as we passed a polite interval before scooting out, I explained how I was racing Peter's father for the Nobel Prize.

Not until the middle of the next week, however, did a nontrivial idea emerge. It came while I was drawing the fused rings of adenine on paper. Suddenly I realized the potentially profound implications of a DNA structure in which the adenine residue formed hydrogen bonds similar to those found in crystals of pure adenine. If DNA was like this, each adenine residue would form two hydrogen bonds to an adenine residue related to it by a 180-degree rotation. Most important, two symmetrical hydrogen bonds could also hold together pairs of guanine, cytosine, or thymine. I thus started wondering whether each DNA molecule consisted of two chains with identical base sequences held together by hydrogen bonds between pairs of identical bases. There was the complication, however, that such a structure could not have a regular backbone, since the purines (adenine and guanine) and the pyrimidines (thymine and cytosine) have different shapes. The resulting backbone would have to show minor in-and-out buckles depending upon whether pairs of purines or pyrimidines were in the center.

Despite the messy backbone, my pulse began to race. If this was DNA, I should create a bombshell by announcing its discovery. The existence of two intertwined chains with idential base sequences could not be a chance matter. Instead it would strongly suggest that one chain in each molecule had at some earlier stage served as the template for the synthesis of the other chain. Under this scheme, gene replication starts with the separation of its two identical chains. Then two new daughter strands are made on the two parental templates, thereby forming two DNA modecules identical to the original molecule. Thus, the essential trick of gene replication could come from the requirement that each base in the newly synthesized chain always hydrogen-bonds to an identical base. That night, however, I could not see why the common tautomeric form of guanine would not hydrogen-bond to adenine. Likewise, several other pairing mistakes should also occur. But since there was no reason to rule out the participation of specific enzymes, I saw no need to be unduly disturbed. For example, there might exist an enzyme specific for adenine that caused adenine always to be inserted opposite an adenine residue on the template strands.

As the clock went past midnight I was becoming more and more pleased. There had been far too many days when Francis and I worried that the DNA structure might turn out to be superficially very dull, suggesting nothing about either its replication or its function in controlling cell biochemistry. But now, to my delight and amazement, the answer was turning out to be profoundly interesting. For over two hours I happily lay awake with pairs of adenine residues whirling in front of my closed eyes. Only for brief moments did the fear shoot through me that an idea this good could be wrong.

My scheme was torn to shreds by the following noon. Against me was the awkward chemical fact that I had chosen the wrong tautomeric forms of guanine and thymine. Before the disturbing truth came out, I had eaten a hurried breakfast at the Whim, then momentarily gone back to Clare to reply to a letter from Max Delbrück which reported that my manuscript on bacterial genetics looked unsound to the Cal Tech geneticists. Nevertheless, he would accede to my request that he send it to the *Proceedings of the National Academy*. In this way, I would still be young when I committed the folly of publishing a silly idea. Then I could sober up before my career was permanently fixed on a reckless course.

At first this message had its desired unsettling effect. But now, with my spirits soaring on the possibility that I had the self-duplicating structure, I reiterated my faith that I knew what happened when bacteria mated. Moreover, I could not refrain from adding a sentence saying that I had just devised a beautiful DNA structure which was completely different from Pauling's. For a few seconds I considered giving some details of what I was up to, but since I was in a rush I decided not to, quickly dropped the letter in the box, and dashed off to the lab.

The letter was not in the post for more than an hour before I knew that my claim was nonsense. I no sooner got to the office and began explaining my scheme than the American crystallographer

Jerry Donohue protested that the idea would not work. The tautomeric forms I had copied out of Davidson's book were, in Jerry's opinion, incorrectly assigned. My immediate retort that several other texts also pictured guanine and thymine in the enol form cut no ice with Jerry. Happily he let out that for years organic chemists had been arbitrarily favoring particular tautomeric forms over their alternatives on only the flimsiest of grounds. In fact, organic-chemistry textbooks were littered with pictures of highly improbable tautomeric forms. The guanine picture I was thrusting toward his face was almost certainly bogus. All his chemical intuition told him that it would occur in the keto form. He was just as sure that thymine was also wrongly assigned an enol configuration. Again he strongly favored the keto alternative.

Jerry, however, did not give a foolproof reason for preferring the keto forms. He admitted that only one crystal structure bore on the problem. This was diketopiperazine, whose three-dimensional configuration had been carefully worked out in Pauling's lab several years before. Here there was no doubt that the keto form, not the enol, was present. Moreover, he felt sure that the quantum-mechanical arguments which showed why diketopiperazine has the keto form should also hold for guanine and thymine. I was thus firmly urged not to waste more time with my harebrained scheme.

Though my immediate reaction was to hope that Jerry was blowing hot air, I did not dismiss his criticism. Next to Linus himself, Jerry knew more about hydrogen bonds than anyone else in the world. Since for many years he had worked at Cal Tech on the crystal structures of small organic molecules, I couldn't kid myself that he did not grasp our problem. During the six months that he occupied a desk in our office, I had never heard him shooting off his mouth on subjects about which he knew nothing.

Thoroughly worried, I went back to my desk hoping that some gimmick might emerge to salvage the like-with-like idea. But it was obvious that the new assignments were its death blow. Shifting the hydrogen atoms to their keto locations made the size differences between the purines and pyrimidines even more important than would be the case if the enol forms existed. Only by the most special pleading could I imagine the polynucleotide backbone bending enough to accommodate irregular base sequences. Even this possibility vanished when Francis came in. He immediately realized that a like-with-like structure would give a 34 Å crystallographic repeat only if each chain had a complete rotation every 68 Å. But this would mean that the rotation angle between successive bases would be only 18 degrees, a value Francis believed was absolutely ruled out by his recent fiddling with the models. Also Francis did not like the fact that the structure gave no explanation for the Chargaff rules (adenine equals thymine, guanine equals cytosine). I, however, maintained my lukewarm response to Chargaff's data. So I welcomed the arrival of lunchtime, when Francis' cheerful prattle temporarily shifted my thoughts to why undergraduates could not satisfy *au pair* girls.

After lunch I was not anxious to return to work, for I was afraid that in trying to fit the keto forms into some new scheme I would run into a stone wall and have to face the fact that no regular hydrogen-bonding scheme was compatible with the X-ray evidence. As long as I remained outside gazing at the crocuses, hope could be maintained that some pretty base arrangement would fall out. Fortunately, when we walked upstairs, I found that I had an excuse to put off the crucial model-building step for at least several more hours. The metal purine and pyrimidine models, needed for systematically checking all the conceivable hydrogen-bonding possibilities, had not been finished on time. At least two more days were needed before they would be in our hands. This was much too long even for me to remain in limbo, so I spent the rest of the afternoon cutting accurate representations of the bases out of stiff cardboard. But by the time they were ready I realized that the answer must be put off till the next day. After dinner I was to join a group from Pop's at the theater.

When I got to our still empty office the following morning, I quickly cleared away the papers from my desk top so that I would have a large, flat surface on which to form pairs of bases held together by hydrogen bonds. Though I initially went back to my like-with-like prejudices, I saw all too well that they led nowhere. When Jerry came in I looked up, saw that it was not Francis, and began shifting the bases in and out of various other pairing possibilities. Suddenly I became aware that an adenine-thymine pair held together by two hydrogen bonds was identical in shape to a guanine-cytosine pair held together by at least two hydrogen bonds. All the hydrogen bonds seemed to form naturally; no fudging was required to make the two types of base pairs identical in shape. Quickly I called Jerry over to ask him whether this time he had any objection to my new base pairs.

When he said no, my morale skyrocketed, for I suspected that we now had the answer to the riddle of why the number of purine residues exactly equaled the number of pyrimidine residues. Two irregular sequences of bases could be regularly packed in the center of a helix if a purine always hydrogen-bonded to a pyrimidine. Furthermore, the hydrogen-bonding requirement meant that adenine would always pair with thymine, while guanine could pair only with cytosine. Chargaff's rules then suddenly stood out as a consequence of a double-helical structure for DNA. Even more exciting, this type of double helix suggested a replication scheme much more satisfactory than my briefly considered like-with-like pairing. Always pairing adenine with thymine and guanine with cytosine meant that the base sequences of the two intertwined chains were complementary to each other. Given the base sequence of one chain, that of its partner was automatically determined. Conceptually, it was thus very easy to visualize how a single chain could be the template for the synthesis of a chain with the complementary sequence.

Upon his arrival Francis did not get more than halfway through the door before I let loose that the answer to everything was in our hands. Though as a matter of principle he maintained skepticism for a few moments, the similarly shaped A-T and G-C pairs had their expected impact. His quickly pushing the bases together in a number of different ways did not reveal any other way to satisfy Chargaff's rules. A few minutes later he spotted the fact that the two glycosidic bonds (joining base and sugar) of each base pair were systematically related by a diad axis perpendicular to the helical axis. Thus, both pairs could be flipflopped over and still have their glycosidic bonds facing in the same direction. This had the important consequence that a given chain could contain both purines and pyrimidines. At the same time, it strongly suggested that the backbones of the two chains must run in opposite directions.

The question then became whether the A-T and G-C base pairs would easily fit the backbone configuration devised during the previous two weeks. At first glance this looked like a good bet, since I had left free in the center a large vacant area for the bases. However, we both knew that we would not be home until a complete model was built in which all the stereochemical contacts were satisfactory. There was also the obvious fact that the implications of its existence were far too important to risk crying wolf. Thus I felt slightly queasy when at lunch Francis winged into the Eagle to tell everyone within hearing distance that we had found the secret of life.

QUESTIONS AND SUGGESTIONS

1. Identify the stylistic devices Watson employs in describing how his scientific discoveries are made.

2. Summarize the process Watson undertook in unlocking the DNA code.

3. You may not be able to write about work that led to the Nobel Prize. But describe a process by which you came to understand something in science that you had not understood before.

RACHEL CARSON

The Obligation to Endure

RACHEL CARSON (1907–1964) was a scientist with the United States Fish and
Wildlife Service. Prior to *Silent Spring,* which appeared in 1962, she had written, among other works, *The Sea Around Us.* However, it was the publication
of *Silent Spring* which created an uproar. Her indictment of the harmful use
of chemical pesticides was attacked as scientifically unsound and emotionally
motivated, even though many who took this position had misunderstood hers.
But without question, her work prompted the federal government to take action against water and air pollution, as well as against persistent pesticides.
    Notice how she combines the techniques of imaginative fiction and scien-
tific exposition. She had struggled with the problem of an overwhelming volume of factual material. Her answer? Rewriting and rewriting.

The history of life on earth has been a history of interaction between living things and their surroundings. To a large extent, the
physical form and the habits of the earth's vegetation and its animal life have been molded by the environment. Considering the
whole span of earthly time, the opposite effect, in which life actually modifies its surroundings, has been relatively slight. Only
within the moment of time represented by the present century
has one species—man—acquired significant power to alter the nature of his world.

During the past quarter century this power has not only increased to one of disturbing magnitude but it has changed in character. The most alarming of all man's assaults upon the environment is the contamination of air, earth, rivers, and sea with
dangerous and even lethal materials. This pollution is for the most
part irrecoverable; the chain of evil it initiates not only in the
world that must support life but in living tissues is for the most
part irreversible. In this now universal contamination of the environment, chemicals are the sinister and little-recognized partners
of radiation in changing the very nature of the world—the very
nature of its life. Strontium 90, released through nuclear explosions into the air, comes to earth in rain or drifts down as fallout,
lodges in soil, enters into the grass or corn or wheat grown there,

and in time takes up its abode in the bones of a human being, there to remain until his death. Similarly, chemicals sprayed on croplands or forests or gardens lie long in soil, entering into living organisms, passing from one to another in a chain of poisoning and death. Or they pass mysteriously by underground streams until they emerge and, through the alchemy of air and sunlight, combine into new forms that kill vegetation, sicken cattle, and work unknown harm on those who drink from once-pure wells. As Albert Schweitzer has said, "Man can hardly even recognize the devils of his own creation."

It took hundreds of millions of years to produce the life that now inhabits the earth—eons of time in which that developing and evolving and diversifying life reached a state of adjustment and balance with its surroundings. The environment, rigorously shaping and directing the life it supported, contained elements that were hostile as well as supporting. Certain rocks gave out dangerous radiation; even within the light of the sun, from which all life draws its energy, there were short-wave radiations with power to injure. Given time—time not in years but in millennia—life adjusts, and a balance has been reached. For time is the essential ingredient; but in the modern world there is no time.

The rapidity of change and the speed with which new situations are created follow the impetuous and heedless pace of man rather than the deliberate pace of nature. Radiation is no longer merely the background radiation of rocks, the bombardment of cosmic rays, the ultraviolet of the sun that have existed before there was any life on earth; radiation is now the unnatural creation of man's tampering with the atom. The chemicals to which life is asked to make its adjustment are no longer merely the calcium and silica and copper and all the rest of the minerals washed out of the rocks and carried in rivers to the sea; they are the synthetic creations of man's inventive mind, brewed in his laboratories, and having no counterparts in nature.

To adjust to these chemicals would require time on the scale that is nature's; it would require not merely the years of a man's life but the life of generations. And even this, were it by some miracle possible, would be futile, for the new chemicals come from our laboratories in an endless stream; almost five hundred annually find their way into actual use in the United States alone.

The figure is staggering and its implications are not easily grasped —500 new chemicals to which the bodies of men and animals are required somehow to adapt each year, chemicals totally outside the limits of biologic experience.

Among them are many that are used in man's war against nature. Since the mid-1940's over 200 basic chemicals have been created for use in killing insects, weeds, rodents, and other organisms described in the modern vernacular as "pests"; and they are sold under several thousand different brand names.

These sprays, dusts, and aerosols are now applied almost universally to farms, gardens, forests, and homes—nonselective chemicals that have the power to kill every insect, the "good" and the "bad," to still the song of birds and the leaping of fish in the streams, to coat the leaves with a deadly film, and to linger on in soil—all this though the intended target may be only a few weeds or insects. Can anyone believe it is possible to lay down such a barrage of poisons on the surface of the earth without making it unfit for all life? They should not be called "insecticides," but "biocides."

The whole process of spraying seems caught up in an endless spiral. Since DDT was released for civilian use, a process of escalation has been going on in which ever more toxic materials must be found. This has happened because insects, in a triumphant vindication of Darwin's principle of the survival of the fittest, have evolved super races immune to the particular insecticide used, hence a deadlier one has always to be developed—and then a deadlier one than that. It has happened also because, for reasons to be described later, destructive insects often undergo a "flareback," or resurgence, after spraying, in numbers greater than before. Thus the chemical war is never won, and all life is caught in its violent crossfire.

Along with the possibility of the extinction of mankind by nuclear war, the central problem of our age has therefore become the contamination of man's total environment with such substances of incredible potential for harm—substances that accumulate in the tissues of plants and animals and even penetrate the germ cells to shatter or alter the very material of heredity upon which the shape of the future depends.

Some would-be architects of our future look toward a time when it will be possible to alter the human germ plasm by design. But we may easily be doing so now by inadvertence, for many chemicals, like radiation, bring about gene mutations. It is ironic to think that man might determine his own future by something so seemingly trivial as the choice of an insect spray.

All this has been risked—for what? Future historians may well be amazed by our distorted sense of proportion. How could intelligent beings seek to control a few unwanted species by a method that contaminated the entire environment and brought the threat of disease and death even to their own kind? Yet this is precisely what we have done. We have done it, moreover, for reasons that collapse the moment we examine them. We are told that the enormous and expanding use of pesticides is necessary to maintain farm production. Yet is our real problem not one of *overproduction?* Our farms, despite measures to remove acreages from production and to pay farmers *not* to produce, have yielded such a staggering excess of crops that the American taxpayer in 1962 is paying out more than one billion dollars a year as the total carrying cost of the surplus-food storage program. And is the situation helped when one branch of the Agriculture Department tries to reduce production while another states, as it did in 1958, "It is believed generally that reduction of crop acreages under provisions of the Soil Bank will stimulate interest in use of chemicals to obtain maximum production on the land retained in crops."

All this is not to say there is no insect problem and no need of control. I am saying, rather, that control must be geared to realities, not to mythical situations, and that the methods employed must be such that they do not destroy us along with the insects.

The problem whose attempted solution has brought such a train of disaster in its wake is an accompaniment of our modern way of life. Long before the age of man, insects inhabited the earth—a group of extraordinarily varied and adaptable beings. Over the course of time since man's advent, a small percentage of the more than half a million species of insects have come into conflict with human welfare in two principal ways: as competitors for the food supply and as carriers of human disease.

Disease-carrying insects become important where human beings are crowded together, especially under conditions where sanitation is poor, as in time of natural disaster or war or in situations of extreme poverty and deprivation. Then control of some sort becomes necessary. It is a sobering fact, however, as we shall presently see, that the method of massive chemical control has had only limited success, and also threatens to worsen the very conditions it is intended to curb.

Under primitive agricultural conditions the farmer had few insect probelms. These arose with the intensification of agriculture—the devotion of immensc acreages to a single crop. Such a system set the stage for explosive increases in specific insect populations. Single-crop farming does not take advantage of the principles by which nature works; it is agriculture as an engineer might conceive it to be. Nature has introduced great variety into the landscape, but man has displayed a passion for simplifying it. Thus he undoes the built-in checks and balances by which nature holds the species within bounds. One important natural check is a limit on the amount of suitable habitat for each species. Obviously then, an insect that lives on wheat can build up its population to much higher levels on a farm devoted to wheat than on one in which whcat is intermingled with other crops to which the insect is not adapted.

The same thing happens in other situations. A generation or more ago, the towns of large areas of the United States lined their streets with the noble elm tree. Now the beauty they hopefully created is threatened with complete destruction as disease sweeps through the elms, carried by a beetle that would have only limited chance to build up large populations and to spread from tree to tree if the elms were only occasional trees in a richly diversified planting.

Another factor in the modern insect problem is one that must be viewed against a background of geologic and human history: the spreading of thousands of different kinds of organisms from their native homes to invade new territories. This worldwide migration has been studied and graphically described by the British ecologist Charles Elton in his recent book *The Ecology of Invasions*. During the Cretaceous Period, some hundred million years ago, flooding seas cut many land bridges between continents and living things found themselves confined in what Elton calls "colossal separate nature reserves." There, isolated from others of their kind, they developed many new species. When some of the land masses were joined again, about 15 million years ago, these species began to move out into new territories—a movement that is not only still in progress but is now receiving considerable assistance from man.

The importation of plants is the primary agent in the modern spread of species, for animals have almost invariably gone along with the plants, quarantine being a comparatively recent and not completely effective innovation. The United States Office of Plant Introduction alone has introduced almost 200,000 species and varieties of plants from all over the world. Nearly half of the 180 or so major insect enemies of plants in the United States are accidental imports from abroad, and most of them have come as hitchhikers on plants.

In new territory, out of reach of the restraining hand of the natural enemies that kept down its numbers in its native land, an invading plant or animal is able to become enormously abundant. Thus it is no accident that our most troublesome insects are introduced species.

These invasions, both the naturally occurring and those dependent on human assistance, are likely to continue indefinitely. Quarantine and massive chemical campaigns are only extremely expensive ways of buying time. We are faced, according to Dr. Elton, "with a life-and-death need not just to find new technological means of suppressing this plant or that animal"; instead we need the basic knowledge of animal populations and their relations to their surroundings that will "promote an even balance and damp down the explosive power of outbreaks and new invasions."

Much of the necessary knowledge is now available but we do not use it. We train ecologists in our universities and even employ them in our governmental agencies but we seldom take their advice. We allow the chemical death rain to fall as though there were no alternative, whereas in fact there are many, and our ingenuity could soon discover many more if given opportunity.

Have we fallen into a mesmerized state that makes us accept as inevitable that which is inferior or detrimental, as though having lost the will or the vision to demand that which is good? Such thinking, in the words of the ecologist Paul Shepard, "idealizes life with only its head out of water, inches above the limits of toleration of the corruption of its own environment . . . Why should we tolerate a diet of weak poisons, a home in insipid surroundings, a circle of acquaintances who are not quite our enemies, the noise of motors with just enough relief to prevent insanity? Who would want to live in a world which is just not quite fatal?"

Yet such a world is pressed upon us. The crusade to create a chemically sterile, insect-free world seems to have engendered a fanatic zeal on the part of many specialists and most of the so-called control agencies. On every hand there is evidence that those engaged in spraying operations exercise a ruthless power. "The regulatory entomologists . . . function as prosecutor, judge and jury, tax assessor and collector and sheriff to enforce their own orders," said Connecticut entomologist Neely Turner. The most flagrant abuses go unchecked in both state and federal agencies.

It is not my contention that chemical insecticides must never be used. I do not contend that we have put poisonous and biologically potent chemicals indiscriminately into the hands of persons largely or wholly ignorant of their potentials for harm. We have subjected enormous numbers of people to contact with these poisons, without their consent and often without their knowledge. If the Bill of Rights contains no guarantee that a citizen shall be secure against lethal poisons distributed either by private individuals or by public officials, it is surely only because our forefathers, despite their considerable wisdom and foresight, could conceive of no such problem.

I contend, furthermore, that we have allowed these chemicals to be used with little or no advance investigation of their effect on soil, water, wildlife, and man himself. Future generations are unlikely to condone our lack of prudent concern for the integrity of the natural world that supports all life.

There is still very limited awareness of the nature of the threat. This is an era of specialists, each of whom sees his own problem and is unaware of or intolerant of the larger frame into which it fits. It is also an era dominated by industry, in which the right to make a dollar at whatever cost is seldom challenged. When the public protests, confronted with some obvious evidence of damaging results of pesticide applications, it is fed little tranquilizing pills of half truth. We urgently need an end to these false assurances, to the sugar coating of unpalatable facts. It is the public that is being asked to assume the risks that the insect controllers calculate. The public must decide whether it wishes to continue on the present road, and it can do so only when in full possession of the facts. In the words of Jean Rostand, "The obligation to endure gives us the right to know."

QUESTIONS AND SUGGESTIONS

1. *Silent Spring* is an attempt to persuade. What distinguishes Carson's enumeration and classification of toxic chemicals from other enumerations in this anthology, such as Charles Darwin's of the types of coral reefs? Where does she use value-laden language; where value-neutral language?

2. Write an essay attempting to persuade an audience that a particular course of action is the right one, using an argument that can be based in scientific facts.

3. Although published in 1962, Carson's theories on progress and preservation are still relevant. Defend this premise with some recent instances of conflict between forces urging land development and those supporting preservation.

NORBERT WIENER

Moral Problems of a Scientist.
The Atomic Bomb. 1942-

NORBERT WIENER (1894–1964) was a mathematician whose interests were
very broad. A prodigy as a child, he was reading Dante and Darwin at the age
of seven, graduated from Tufts University at 14, and received a Ph.D. in philosophy from Harvard at 18. He tells the story of his life in *Ex-Prodigy*, and
continues it in *I Am a Mathematician*, from which this selection is taken.
  He pioneered in the field of information theory, coining the term "cybernetics." Toward the end of his life, he had become interested in the application of concepts from information theory to problems in biology. From 1919,
he was a well-loved and increasingly legendary member of the Massachusetts
Institute of Technology faculty.

One day during the Second World War I was called down to
Washington to see Vannevar Bush. He told me that Harold Urey,
of Columbia, wanted to see me in connection with a diffusion
problem that had to do with the separation of uranium isotopes.
We were already aware that uranium isotopes might play an important part in the transmutation of elements and even in the
possible construction of an atomic bomb, for the earlier stages of
this work had come before the war and had not been made in the
United States.

  I went to New York and had a talk with Urey, but I could not
find that I had any particular qualification for solving the special
problem on which he requested help. I was also very busy with my
own work on predictors. I felt that there I had found my niche for
the duration of the war. It was a place where my own ideas were
particularly useful, and where I did not feel that anyone else could
do quite so good a job without my help.

  I therefore showed no particular enthusiasm for Urey's problem,
although I did not say in so many words that I would not work on
it. Perhaps I was not cleared for the problem, or perhaps my lack
of enthusiasm itself was considered as a sufficient reason for not

using me, but that was the last I heard of the matter. This work was a part of the Manhattan Project and the development of the atomic bomb.

Later on, various young people associated with me were put on the Manhattan Project. They talked to me and to everyone else with a rather disconcerting freedom. At any rate, I gathered that it was their job to solve long chains of differential equations and thereby handle the problem of repeated diffusions. The problem of separating uranium isotopes was reduced to a long chain of diffusions of liquids containing uranium, each stage of which did a minute amount of separation of the two isotopes, ultimately leading in the sum to a fairly complete separation. Such repeated diffusions were necessary to separate two substances as similar in their physical and chemical properties as the uranium isotopes. I then had a suspicion (which I still have, though I know nothing of the detail of the work) that the greater part of this computation was an expensive waste of money. It was explained to me that the effects on which one was working were so vanishingly small that without the greatest possible precision in computation they might have been missed altogether.

This however did not look reasonable to me, because it is exactly under these circumstances of the cumulative use of processes which accomplish very little each time that the standard approximation to a system of differential equations by a single partial differential equation works best. In other words, I had and have the greatest doubts, to the effect that in this very slow and often repeated diffusion process those phenomena which may not be justifiably treated as continuous are of very slight real importance.

Be that as it may, while I did not have any detailed knowledge of what was being done on Manhattan Project, the time came when neither I nor any other active scientist in America could fail to be aware that such a project was under way. Even then we did not have any clear idea of how it was to be used. We were afraid that the main use to be made of radioactive isotopes was as poisons. We feared that here we might well find ourselves in the position of having developed a weapon which international morality and policy would not permit us to use, even as they had held the Germans back from the use of poison gas against cities. Even were the work to result in an explosive, we were not at all clear as to the possibilities of the bomb nor as to the moral problems which its use would involve. I was very certain of at least one thing: that I was most happy to have had no share in the responsibility for its development and its later use.

So far the moral problem of warfare had not concerned me directly. However, in the fall of 1944 a complex of events took place which had a very considerable effect on my later career and thought. I had already begun to reflect on the relation between the high-speed computing machine and the automatic factory, and I had come to the conclusion that as the essence of the computing machine lay in its speed and in its programming, or determination of the sequence of operations to be performed by means of a magnetic tape or punched cards, the automatic factory was not far off. I wondered whether I had not got into a moral situation in which my first duty might be to speak to others concerning material which could be socially harmful.

The automatic factory could not fail to raise new social problems concerning employment, and I was not sure that I had the answers. A vast redistribution of labor at different levels would be created. When the human being is being used mechanically, simply as an inferior sort of switching or decision device, the automatic factory threatens to replace him completely by mechanical agencies. On the other hand, it creates a new demand for the highly skillful professional man who can organize the order of operations which will best serve a particular function.

It will also create a demand for trouble-shooters and maintenance crews of a particularly well-trained sort. If these changes in the demand for labor come upon us in a haphazard and ill-organized way, we may well be in for the greatest period of unemployment we have yet seen. It seemed to me then quite possible that we could avoid a catastrophe of this sort, but if so, it would only be by much thinking, and not by waiting supinely until the catastrophe is upon us. I shall say something later in this chapter of my present opinion in the matter.

Accordingly, when a colleague wished for some information concerning my newer work I answered that I was not by any means certain that this work should be communicated to him, or indeed to the public at large. I felt all the more strongly about this inasmuch as he had requested the information for military purposes, and I did not know whether I should be a party to the use of my new ideas for controlled missiles and the like.

I showed the letter to a colleague of mine who happened to have a flair for journalism. He immediately suggested that I send my note to the *Atlantic Monthly* as a basis for what might be a more elaborate article. I followed his suggestion and sent it. If I had thought out fully how I was thus subjecting myself to a deep moral commitment while he was subjecting himself to nothing at all, I might well have hesitated, although I probably would put this hesitation behind me as an act of cowardice. The moral consequences of my act were soon to follow.

About this time I had agreed to participate in two meetings: the earlier one on applied mathematics, called together by Princeton at the end of its second hundred years of existence; and a later one organized by Aiken at Harvard on the subject of automatic high-speed computing machines. To this second meeting, which took place under the joint auspices of Harvard University and the Navy Department, I had agreed to give a paper. Meanwhile I reported at the Princeton meeting on my work on prediction theory. While my talk covered material which I knew could ultimately be used for military purposes, I had counted on the abstractness of my presentation and the natural inertia of many of my colleagues to prevent the work from being put to immediate and uncontrollable military use.

However, my hand was forced. A colleague who had previously taught at M.I.T. and had gone back to his home University of California, because of his earlier associations there and because of the climate, had been pushing my name as head of or consultant for a military or semimilitary project on mechanical computation, to be located in California. He had not consulted me in the matter, but he had assumed that an invitation to California would be accepted by me without question.

I have said that the project was semimilitary. It was in fact to be under the Bureau of Standards, but it was quite clear that all the facilities engineered by the project if it should be successful would be pre-empted for years by the military services. My acquaintance had tried to commit me not only to work whose objective was distasteful to me but to work which would involve conditions of secrecy, of a police examination of my opinions, and of the confinements of administrative responsibility. These I could not accept. They would have bound me to a course of conduct which would have broken me down in a very few months. When the invitation was passed on to me, I considered my *Atlantic Monthly* letter, and I had no alternative but to say no. I probably would have said no anyway, as I did not fancy myself as an administrator.

Then I recalled the military meeting at Harvard at which I had already committed myself to speak. There were some two weeks before it would come off, so I thought that I had ample time to change my policy. I went to Aiken and tried to explain the situation. In particular, I pointed out to him that the California offer had made it necessary for me to take a definite stand on my war work and that I could not accept one sort of a military association and reject another. I therefore asked to be released from my promise to give a talk.

I gathered from Aiken that there would be time to take my name off the list of speakers for the meeting. However, when the meeting came, I found that Aiken had done this by merely running a line through my name on the printed programs which had been issued to members of the meeting and to the press.

This procedure was extremely embarrassing to me. It was even more embarrassing to him. The newspapermen came to me and asked whether this striking out of my name had anything to do with the letter that had appeared in the *Atlantic Monthly*. I said that it had, and I tried to explain to them the circumstances that had forced my hand. I took full responsibility in the matter and said that I acted with Aiken's consent and that I was not taking this step out of pique or personal animosity.

Naturally, they went to Aiken in the matter. Without reflection, he assumed that I had been involved in some deep plot to discredit him and to turn the meeting into a public scandal. In fact, as I have said, I had consulted him at the very beginning and had understood that there was to be no publicity about the affair. The matter would have had no publicity if it had not been for the emphasis which he had placed on my participation in the meeting by the way he scratched out my name.

All these emotional experiences were nothing to those through which I went at the time of the bombing of Hiroshima. At first I was of course startled, but not surprised, as I had been aware of the possibility of the use of the new Manhattan Project weapons against an enemy. Frankly, however, I had been clinging to the hope that at the last minute something in the atomic bomb would fail to work, for I had already reflected considerably on the significance of the bomb and on the meaning to society of being compelled to live from that time on under the shadow of the threat of limitless destruction.

Of course I was gratified when the Japanese war ended without the heavy casualties on our part that a frontal attack on the mainland would have involved. Yet even this gratifying news left me in a state of profound disquiet. I knew very well the tendency (which is not confined to America, though it is extremely strong here) to regard a war in the light of a glorified football game, at which at some period the final score is in, and which we have to count as either a definite victory or a definite defeat. I knew that this attitude of dividing history into separate blocks, each contained within itself, is by no means weakest in the Army and Navy.

But to me this episodic view of history seemed completely superficial. The most important thing about the atomic bomb was, in my opinion, not the termination of a specific war without undue casualties on our part, but the fact that we were now confronted with a new world and new possibilities with which we should have to live ever after. To me the most important fact about the wars of the past was that, serious as they had been, and completely destructive for those involved in them, they had been more or less local affairs. One country and one civilization might go under, but the malignant process of destruction had so far been localized, and new races and peoples might take up the torch which the others had put down.

I did not in the least underrate the will to destructiveness, which was as much a part of war with a flint ax and of war with a bow and arrow as it is of war with a musket and of war with a machine gun. What came most strongly to my attention was that in previous wars the power of destruction was not commensurate with the will for destruction. Thus, while I realized that as far as the people killed or wounded are concerned, there is very little difference between a cannonade or an aerial bombardment with explosive bombs of the type already familiar and the use of the

atomic bomb, there seemed to me to be most important practical
differences in the consequences to humanity at large.

Up to now no great war, and this includes World War II, had
been possible except by the concerted and prolonged will of the
people fighting, and consequently no such war could be undertaken without a profoundly real share in it by millions of people.
Now the new modes of mass destruction, expensive as they must
be in the bulk, have become so inexpensive per person killed that
they no longer take up an overwhelming share of a national
budget.

For the first time in history, it has become possible for a limited
group of a few thousand people to threaten the absolute destruction of millions, and this without any highly specific immediate
risk to themselves.

War had made the transition between an overwhelming assertion of national effort and the push-button declaration of the will
of a small minority of people. Fundamentally this is true, even if
one includes in the military effort all the absolutely vast but relatively small sums which have been put into the whole body of
nuclear research. It is even more devastatingly true if one con-
siders the relatively minimal effort required on the part of a few
generals and a few aviators to place on a target an atomic bomb
already made.

Thus, war has been transported, at least as a possibility, from
the field of national effort to the field of private conspiracy. In
view of the fact that the great struggle to come threatens to be one
between the United States and the Soviet Government, and in
view of the additional fact that the whole atmosphere and administration of the Soviet Government shares with that of the Nazis an
extremely strong conspiratorial nature, we have taken a step which
is intrinsically most dangerous for us.

I did not regard with much seriousness the assertions which
some of the great administrators of science were making, to the
effect that the know-how needed for the construction of the atomic
bomb was a purely American thing and could not be duplicated
by a possible enemy for many years at least, during which we
could be counted upon to develop a new and even more devastat-
ing know-how. In the first place, I was acquainted with more than
one of these popes and cardinals of applied science, and I knew

very well how they underrated aliens of all sorts, in particular those not of the European race. With my wide acquaintance among scholars of many races and many countries, I had not been able to discern that scientific ability and moral discipline were the peculiar property of those of blanched skin and English speech.

But this was not all. The moment that we had declared both our possession of the bomb and its efficiency by using it against an enemy, we had served notice on every country that its continued existence and independence of policy were conditioned on its prompt possession of a similar weapon. This meant two things: that any country which was our rival or potential rival was bound to push nuclear research for the sake of its own continued independent existence, with the greatly stimulating knowledge that this research was not intrinsically in vain; and that any such country would inevitably set up an espionage system to get hold of our secrets.

This is not to say that we Americans would not be bound in self-defense to oppose such leaks and such espionage with our full effort for the sake of our very national existence; but it does mean that such considerations of legality and such demands on the moral responsibility of loyal American citizens could not be expected to have the least force beyond our frontiers. If the roles of Russia and the United States had been reversed, we should have been compelled to do exactly what they did in attempting to discover and develop such a vital secret of the other side; and we should regard as a national hero any person attached to our interests who performed an act of espionage exactly like that of Fuchs or the Rosenbergs.

I then began to evolve in my mind the general problem of secrets; not so much as a moral issue, but as a practical issue and a policy which we might hope to maintain effectively in the long run. Here I could not help considering how soldiers themselves regard secrets in the field. It is well recognized that every cipher can be broken if there is sufficient inducement to do it and if it is worthwhile to work long enough; and an army in the field has one-hour ciphers, twenty-four-hour ciphers, one-week ciphers, perhaps one-year ciphers, but never ciphers which are expected to last an eternity.

Under the ordinary circumstances of life, we have not been accustomed to think in terms of espionage, cheating, and the like. In particular, such ideas are foreign to the nature of the true scientist, who, as Einstein has pointed out, has as his antagonist a world which is hard to understand and interpret, but which does not maliciously and malignantly resist this interpretation. "The Lord is subtle, but he isn't plain mean."

With ordinary secrets of limited value, we do not have to live under a perpetual fear that somebody is trying to break them. If, however, we establish a secret of the supreme value and danger of the atomic bomb, it is not realistic to suppose that it will never be broken, nor that the general good will among scientists will exclude the existence of one or two who, either because of their opinions or their slight resistance to moral pressure, may give our secrets over to those who will endanger us.

If we are to play with the edged tools of modern warfare, we are running not merely the danger of being cut by accident and carelessness but the practical certainty that other people will follow where we have already gone and that we shall be exposed to the same perils to which we have exposed others. Secrecy is thus at once very necessary and, in the long run, quite impossible. It is unrealistic to give over our main protection to such a fragile defense.

There were other reasons, moreover, which, on much more specific grounds, made me feel skeptical of the wisdom of the course we had been pursuing. It is true that the atomic bomb had been perfected only after Germany had been eliminated from the war, and that Japan was the only possible proving ground for the bomb as an actual deadly weapon. Nevertheless, there were many both in Japan and elsewhere in the Orient who would think that we had been willing to use a weapon of this terribleness against Japan when we might not have been willing to use it against a white enemy. I myself could not help wondering whether there might not be a certain degree of truth in this charge. In a world in which European colonialism in the Orient was rapidly coming to an end, and in which every oriental country had much reason to be aware of the moral difference which certain elements in the West were in the habit of making between white people and colored people, this weapon was pure dynamite (an obsolete metaphor now that the atomic bomb is here) as far as our future diplomatic policy was concerned. What made the situation ten times worse was that this was the sort of dynamite which Russia, our greatest potential antagonist if not our greatest actual enemy, was in a position to use, and would have no hesitation whatever in using.

It is the plainest history that our atomic bomb effort was international in the last degree and was made possible by a group of people who could not have been got together had it not been for the fact that the threat of Nazi Germany was so strongly felt over the world, and particularly by that very scholarly group who contributed the most to Nuclear theory. I refer to such men as Einstein, Szilard, Fermi, and Niels Bohr. To expect in the future that a similar group could be got together from all the corners of the world to defend our national policy involved the continued expectation that we should always have the same moral prestige. It was therefore doubly unfortunate that we should have used the bomb on an occasion on which it might have been thought that we would not have used it against white men.

There was another matter which aroused grave suspicion in the minds of many of us. While the nuclear program did not itself involve any overwhelming part of the national military effort, it was still in and for itself an extremely expensive business. The people in charge of it had in their hands the expenditure of billions of dollars, and sooner or later, after the war, a day of reckoning was bound to come, when Congress would ask for a strict accounting and for a justification of these enormous expenditures. Under these circumstances, the position of the high administrators of nuclear research would be much stronger if they could make a legitimate or plausible claim that this research had served a major purpose in terminating the war. On the other hand, if they had come back empty-handed—with the bomb still on the docket for future wars, or even with the purely symbolic use of the bomb to declare to the Japanese our willingness to use it in actual fact if the war were to go on—their position would have been much weaker, and they would have been in serious danger of being broken by a new administration coming into power on the rebound after the war and desirous of showing up the graft and ineptitude of its predecessor.

Thus, the pressure to use the bomb, with its full killing power, was not merely great from a patriotic point of view but was quite as great from the point of view of the personal fortunes of people involved in its development. This pressure might have been unavoidable, but the possibility of this pressure, and of our being forced by personal interests into a policy that might not be to our best interest, should have been considered more seriously from the very beginning.

Of the splendid technical work that was done in the construction of the bomb there can be no question. Frankly, I can see no evidence of a similar high quality of work in the policy-making which should have accompanied this. The period between the experimental explosion at Los Alamos and the use of the bomb in deadly earnest was so short as to preclude the possibility of clear thinking. The qualms of the scientists who knew the most about what the bomb could do, and who had the clearest basis to estimate the possibilities of future bombs, were utterly ignored, and the suggestion to invite Japanese authorities to an experimental exhibition of the bomb somewhere in the South Pacific was flatly rejected.

Behind all this I sensed the desires of the gadgeteer to see the wheels go round. Moreover, the whole idea of push-button warfare has an enormous temptation for those who are confident of their power of invention and have a deep distrust of human beings. I have seen such people and have a very good idea of what makes them tick. It is unfortunate in more than one way that the war and the subsequent uneasy peace have brought them to the front.

All these and yet other ideas passed through my mind on the very day of Hiroshima. One of the strong points and at the same time one of the burdens of the creative scholar is that he must stand alone. I wished—oh how I wished!—that I could be in a position to take what was happening passively, with a sincere acceptance of the wisdom of the policy makers and with an abdication of all personal judgment. The fact is, however, that I had no reason to believe that the judgment of these men on the larger issues of the situation was superior to my own, whatever their technical information might be. I knew that more than one of the high officials of science had not one tenth my contact with the scientists of other countries and of other standpoints and was in nowhere nearly as good a position to assess the world reaction to the bomb. I knew, moreover, that I had been in the habit of considering the history of science and of invention from a more or less philosophic point of view, and I did not believe that those who made the decisions could do this any better than I might. The sincere scientist must back his bets and guesses, even when he is a Cassandra and no one believes him. I had behind me many years of lonely work in science where I had finally proved to be in the right. This inability to trust the Powers That Be was a source of no particular satisfaction to me, but there it was, and it had to be faced.

One of my greatest worries was the reaction of the bomb on science and on the public's attitude to the scientist. We had voluntarily accepted a measure of secrecy and had given up much of our liberty of action for the sake of the war, even though—for that very purpose, as many of us thought—more secrecy than the optimum was imposed, and this at times had hampered our internal communications more than the information-gathering service of the enemy. We had hoped that this unfamiliar self-discipline would be a temporary thing, and we had expected that after this war—as, after all, before—we should return to the free spirit of communication, intranational and international, which is the very life of science. Now we found that, whether we wished it or not, we were to be the custodians of secrets on which the whole national life might depend. At no time in the forseeable future could we again do our research as free men. Those who had gained rank and power over us during the war were most loath to relinquish any part of the prestige they had obtained. Since many of us possessed secrets which could be captured by the enemy and could be used to our national disadvantage, we were obviously doomed to live in an atmosphere of suspicion forever after, and the police scrutiny on our political opinions which began in the war showed no signs of future remission.

The public liked the atomic bomb as little as we did, and there were many who were quick to see the signs of future danger and to develop a profound consciousness of guilt. Such a consciousness looks for a scapegoat. Who could constitute a better scapegoat than the scientists themselves? They had unquestionably developed the potentialities which had led to the bomb. The man in the street, who knew little of scientists and found them a strange and self-contained race, was quick to accuse them of a desire for the power of destruction manifested by the bomb. What made this both more plausible and more dangerous was the fact that, while the working scientists felt very little personal power and had very little desire for it, there was a group of administrative gadget workers who were quite sensible of the fact that they now had a new ace in the hole in the struggle for power.

At any rate, it was perfectly clear to me at the very beginning that we scientists were from now on to be faced by an ambivalent attitude. For the public, who regarded us as medicine men and magicians, was likely to consider us an acceptable sacrifice to the gods as other, more primitive publics do. In that very day of the atomic bomb the whole pattern of the witch hunt of the last eight years became clear, and what we are living through is nothing but the transfer into action of what was then written in the heavens.

While I had no share in the atomic bomb itself, I was nevertheless led into a very deep searching of soul. I have already explained how my work on prediction and on computing machines had led me to the basis of cybernetics, as I was later to call it, and to an understanding of the possibilities of the automatic factory. From the strictly scientific point of view, this was not as revolutionary as the atomic bomb, but its social possibilities for good and for evil were enormous. I tried to see where my duties led me, and if by any chance I ought to exercise a right of personal secrecy parallel to the right of governmental secrecy assumed in high quarters, suppressing my ideas and the work I had done.

After toying with the notion for some time, I came to the conclusion that this was impossible, for the ideas which I possessed belonged to the times rather than to myself. If I had been able to suppress every word of what I had done, they were bound to reappear in the work of other people, very possibly in a form in which the philosophic significance and the social dangers would be stressed less. I could not get off the back of this bronco, so there was nothing for me to do but to ride it.

I thus decided that I would have to turn from a position of the greatest secrecy to a position of the greatest publicity, and bring to the attention of all the possibilities and dangers of the new developments. I first thought of the trade unions as the people who would naturally be most interested in the matter. My friends directed me towards two union leaders, one of them an intellectual counselor who had himself very little direct authority among the union people with whom he was associated, and the other a high official of the typographers' union. In both cases I found a confirmation of what my English friends had told me some years before: The union official comes too directly from the workbench, and is too immediately concerned with the difficult and highly technical problems of shop stewardship, to be able to entertain any very forward-looking considerations of the future of his own craft.

I found plenty of good will among my union friends but an absolute block on their part to communicate my ideas to their union workers. This was in the middle forties; since then the situation has changed radically. I have been in repeated communication with Mr. Walter Reuther, of the United Automobile Workers, and I have found in him both an understanding of my problems and a willingness to give my ideas publicity through his union journals. In fact, I have found in Mr. Reuther and the men about him exactly that more universal union statesmanship which I had missed in my first sporadic attempts to make union contacts.

There is another quarter in which the sort of ideas I have had concerning the automatic factory have made gratifying headway. This is in the circles of management itself. In the winter of 1949 I gave a talk to the Society for the Advancement of Management concerning the automatic factory as a technical possibility and the social problems it would introduce, and in both matters I was backed up by high management authorities, as, for example, an executive of Remington Rand, Inc. In December of 1952 I was asked to give a talk on a similar subject as part of a symposium on the automatic factory held by the American Society of Mechanical Engineers.

The progress in the general attitude from the first talk to the second was remarkable. Not only was the attending public much larger and my technical remarks confirmed by automatic-machine men for several industries, but the social consciousness of the group as a whole was far beyond what I had found three years before.

While there were a good many who were more sanguine than I had been as to the possibility of achieving a large measure of industrial automatization without catastrophe, there was a general awareness of the interest of the public at large in a meeting which was going to affect so profoundly their future method of life. In particular, problems of the grade-up of repetitive factory workers into trouble-shooting men (and indeed into a sort of junior engineer) occupied a great deal of attention.

Another much-debated problem was that of the new leisure we might expect in the future and the use that could and must be made of it. Indeed, I heard hard-boiled engineering administrators express views which sounded remarkably like the writings of William Morris. Above all, I had everyone backing me in cautioning that the new displacement of human beings from the repetitive labor of the factory must not be taken as a devaluation of the human being and a glorification of the gadget.

The years that have passed since this talk have seen the automatic factory develop from a remote possibility into a beginning actuality, and we can start to assess on a factual basis its probable impact on society. The first industrial revolution of the early nineteenth century replaced the individual by the machine as a source of power. No factory worker of the present day is earning any large part of his wages by the horsepower of his output. Even if he is doing the hardest sort of physical labor, as for example, if he is a steel puddler, his pay is not primarily given him as a prime mover in a power process. What he is actually paid for is his experience and knowledge of how to exert his strength most effectively in a highly purposeful manufacturing process.

However, the strong men of industry such as the steel puddlers are in a decided minority. The factory worker finds a small electric motor or a pneumatic tool at his elbow, and these will give him the sheer physical strength of ten men. His business is to accomplish a certain purpose by going through certain motions in a given succession. If, for example, he is pasting labels on tin cans, he must see that he has the correct stack of labels before him, that he has moistened them correctly, that he has put them in the correct position on the can, and that he turns at the proper time from one can to the next. This sort of laborer goes through a purely repetitive process, making the minimum demands on any but the lowest level of judgment and observation.

Of course, there are other forms of factory labor. There are the foremen and there are the members of the trouble-shooting gangs, who at the very lowest level must be skilled craftsmen and, on the higher levels, are assimilated in their function to junior engineers. Leaving out these higher ranks of labor, the routine factory worker is often doing so conventional a task that every motion of his and the cue for every motion, may be assigned in advance. This is the point of such efficiency systems as the motion study of Taylor and the Gilbreths.

I have already indicated that it is this level of work which will be replaced by the operations of the automatic factory. Essentially, to my way of seeing things, most of the human labor which the automatic factory displaces is an inhuman sort of human labor, which has been considered a natural task for human beings only since the historical accident of the industrial revolution. Nevertheless, any sudden and uncompensated displacement of this labor must have catastrophic consequences in the direction of unemployment.

Where will this labor have to go? The most obvious answer is that even the automatic factory will always require a considerable group of trouble shooters, skilled craftsmen, and specialists in programming or in the adaptation of the machines to specific problems. During a gradual process of automatization, the natural place for unskilled factory labor to go is into these higher cadres, by some sort of up-grading. The question of the possibility of this up-grading thus becomes vital.

There is a considerable amount of evidence that the sources of labor which furnished the unskilled factory labor of the past generation are drying up, because, since soon after the end of the First World War, we have had no extensive body of immigrants seeking to establish and settle themselves in the country and willing to accept any degree of economic undervaluing. It is the children of this last extensive generation of immigrants who fought in the Second World War, and the rising generation of the present day consists of their children's children. These younger generations are unwilling to accept the permanent position of economic inferiority belonging to the unskilled workers in the old type of factory. Many of them are going into the professions, and even those who are not are beginning to demand that their work be interesting and not a blind alley.

This is not the first time in our industrial history in which technical advances have been conditioned by the decreasing availability of labor of a certain type. Automatic telephone switching came in simply because the old system of hand switching bade fair to demand the entire population of girl high school graduates.

Another matter which may make the stepping up of labor easier than it might have seemed a few years ago is the training of a very considerable part of the young men in our military services as technicians of a relatively high grade. This has been particularly the case in the Air Force. The sort of young man who can be trained to direct and to care for a radar instrument is certainly the sort who can easily learn to be a member of a factory trouble-shooting gang.

Thus it is quite possible, although it is not certain, that the labor environment for the automatic factory has come just at the right time. At any rate, the atmosphere into which the automatic factory is coming is one where it fits into a definite niche of human activity, and one which has been alerted to both the advantages of automatization and the risks.

While this has been the work of many hands, I feel proud that some part of the healthy and understanding atmosphere into which automatization is coming, and of the collaboration of labor and management in being prepared to work out jointly a mode of industrial life which embraces the automatic factory, may be due to my early efforts to alert both of these elements.

QUESTIONS AND SUGGESTIONS

1. Contrast Wiener's style of scientific autobiography with James Watson's.

2. How far do you believe the scientist's responsibility for the uses of his research extends? Support your case through examples.

less slowly. With the scientific journal (the first one was the Royal Society's *Philosophical Transactions*), scientists began to experiment with forms of communication that suited their new needs. They recorded the progress of experiments; they presented results; they described their findings—microscopic animals in pond water, a new way of growing apples, a strange animal observed by a traveler in Africa. The style of these early reports seems verbose and even casual to us, but, in contrast to the prose of their contemporaries, we can see that these writers valued clear, unadorned, natural expression, a form of writing that they thought imitated the processes of the rational mind operating according to the experimental method. The modern style of scientific writing has come from such roots.

Why was it so important for these scientists to communicate among themselves? For one thing, the acceptance of any theory depended upon the ability of a scientist to persuade the scientific community that it was valid. The name of a Newton or a Darwin seems to stand alone, but both would be forgotten, as Mendel almost was, if they had been unable to convince other scientists—men with their own beliefs and prejudices—of the correctness of the new ways of thinking. Writing well about scientific ideas to an audience not fully acquainted with the particular field or prepared to accept new developments, then, is as important to the scientist as mastery of a field and can be as demanding as creative work in it.

There are other reasons, of course, for the persistence of both formal and informal channels of communication among scientists, among them the need to establish priority of discovery and to further the work of others in the same field. Thus, even though in the early stages of science, a scientist's most original and important work would be published in a book—witness Newton's *Principia*—by the end of the eighteenth century, because it served science's purposes so well, the journal was the place where scientists published. Darwin's and Wallace's first papers on evolution were read before the Linnaean Society and published in its proceedings; a year later, Darwin brought out the *Origin of Species* to a wider audience of scientists and non-scientists alike.

By the end of the nineteenth century, the scientific article had begun to take on its present shape, with footnotes, bibliography, and "objective" style, barely resembling its ancestor, the letter, at all. The vast expansion of scientific research and the age of the computer have led to the abstract, a succinct summary of an article. Computerized data banks, taking key words from abstracts, perform literature searches, putting researchers in touch with work which might be of real value to them. In the past, when scientists were fewer in number, one man, the "intelligencer," performed a very similar function, but his brain was the computer.

Still, much communication among scientists is informal, or on a personal level—professional meetings, private correspondence, papers circulated among colleagues. In *The Double Helix,* James Watson refers to the work of Linus Pauling and of Rosalind Franklin; he knew of their work through informal channels.

There persists the question of jargon in writing for professional audiences. Specialized vocabularies develop when new knowledge like new "territory" is found on the sphere of knowledge and is named. All verbal communication is through codes; a word, when we speak it, calls up a territory in the mind of the hearer, an image, a concept. Jargon, then, is code.

The degree to which such specialized vocabularies are legitimate simply depends upon audience. Hence, scientific journals fall into three groups, each with different "jargon" levels. Some are specialized for scientists in a particular field, such as the *American Journal of Physics;* some are to be read by scientists from many fields, such as *Nature* in Britain and *Science* in America. Some are for professional and general audiences alike, notably *Scientific American,* which has probably been the single most important force for excellence in scientific writing in America. James Watson and Francis Crick's DNA article appeared on the pages of *Nature* just as they had written it; but, had they sent it to *Scientific American,* it would have gone through several editings to make it intelligible to a broader audience. Notice the similarities between the specialized vocabulary of Lynn Margulis's treatment of DNA addressed to a pre-professional audience, and James Watson and Francis Crick's, addressed to a highly select professional audience. These authors have all considered the vocabularies of their audiences carefully and cannot be accused of misusing specialized words to confuse the reader or to conceal inadequacies of thought.

THOMAS JEFFERSON

Instructions to Captain Lewis

Son of Peter Jefferson (surveyor, mapmaker, and magistrate) and Jane Randolph (of a prominent Virginia family), THOMAS JEFFERSON (1743–1826) is best known for drafting the Declaration of Independence and for being thrice elected President of the United States. At the College of William and Mary he excelled in the study of mathematics; upon graduation in 1762 he undertook the study of law with George Wythe and in 1769 began construction of his Monticello estate. He held a variety of positions in the Virginia government, among them representative to Continental Congress. He founded the University of Virginia, and supported the expeditions of Lewis and Clark. Throughout his life, Jefferson pursued an active career as a writer (beyond the necessary correspondences, government documents, and treaties). His *Notes on the State of Virginia* (1787) is a comprehensive history of the state's geography, population, and politics. His *Autobiography* (1821) details his political activities during the Revolutionary period of the 1780s.

June 20, 1805

To Merryweather Lewis, Esq., Captain of the 1st Regiment of Infantry of the United States of America.

Your situation as Secretary of the President of the United States has made you acquainted with the objects of my confidential message of Jan. 18, 1803, to the legislature. You have seen the act they passed, which, tho' expressed in general terms, was meant to sanction those objects, and you are appointed to carry them into execution.

Instruments for ascertaining by celestial observations the geography of the country thro' which you will pass, have been already provided. Light articles for barter, & presents among the Indians, arms for your attendants, say for from 10 to 12 men, boats, tents, & other travelling apparatus, with ammunition, medicine, surgical instruments & provision you will have prepared with such aids as the Secretary at War can yield in his department; & from him also you will receive authority to engage among our troops, by voluntary agreement, the number of attendants above mentioned, over whom you, as their commanding officer are invested with all the powers the laws give in such a case.

As your movements while within the limits of the U.S. will be better directed by occasional communications, adapted to circumstances as they arise, they will not be noticed here. What follows will respect your proceedings after your departure from the U.S.

Your mission has been communicated to the Ministers here from France, Spain, & Great Britain, and through them to their governments: and such assurances given them as to its objects as we trust will satisfy them. The country of Louisiana having been ceded by Spain to France, the passport you have from the Minister of France, the representative of the present sovereign of the country, will be a protection with all its subjects: And that from the Minister of England will entitle you to the friendly aid of any traders of that allegiance with whom you may happen to meet.

The object of your mission is to explore the Missouri river, & such principal stream of it, as, by its course & communication with the water of the Pacific Ocean may offer the most direct & practicable water communication across this continent, for the purposes of commerce.

Beginning at the mouth of the Missouri, you will take observations of latitude and longitude at all remarkable points on the river, & especially at the mouths of rivers, at rapids, at islands & other places & objects distinguished by such natural marks & characters of a durable kind, as that they may with certainty be recognized hereafter. The courses of the river between these points of observation may be supplied by the compass, the log-line & by time, corrected by the observations themselves. The variations of the compass too, in different places should be noticed.

The interesting points of the portage between the heads of the Missouri & the water offering the best communication with the Pacific Ocean should be fixed by observation & the course of that water to the ocean, in the same manner as that of the Missouri.

Your observations are to be taken with great pains & accuracy, to be entered distinctly, & intelligibly for others as well as yourself, to comprehend all the elements necessary, with the aid of the usual tables to fix the latitude & longitude of the places at which they were taken, & are to be rendered to the war office, for the purpose of having the calculations made concurrently by proper persons within the U.S. Several copies of these as well as of your other notes, should be made at leisure times & put into the care of the most trustworthy of your attendants, to guard by multiplying them against the accidental losses to which they will be exposed. A further guard would be that one of these copies be written on the paper of the birch, as less liable to injury from damp than common paper.

The commerce which may be carried on with the people inhabiting the line you will pursue, renders a knowledge of these people important. You will therefore endeavor to make yourself acquainted, as far as a diligent pursuit of your journey shall admit.

with the names of the nations & their numbers;

the extent & limits of their possessions;

their relations with other tribes or nations;

their language, traditions, monuments;

their ordinary occupations in agriculture, fishing, hunting, war, arts, & the implements for these;

their food, clothing, & domestic accommodations;

the diseases prevalent among them, & the remedies they use;

moral and physical circumstance which distinguish them from the tribes they know;

peculiarities in their laws, customs & dispositions;

and articles of commerce they may need or furnish & to what extent.

And considering the interest which every nation has in extending & strengthening the authority of reason & justice among the people around them, it will be useful to acquire what knowledge you can of the state of morality, religion & information among them, as it may better enable those who endeavor to civilize & instruct them, to adapt their measures to the existing notions & practises of those on whom they are to operate.

Other objects worthy of notice will be

the soil & face of the country, its growth & vegetable productions; especially those not of the U.S.

the animals of the country generally, & especially those not known in the U.S.

The remains & accounts of any which may be deemed rare or extinct;

the mineral productions of every kind; but more particularly

Should you reach the Pacific Ocean inform yourself of the circumstances which may decide whether the furs of those parts may not be collected as advantageously at the head of the Missouri (convenient as is supposed to the waters of the Colorado & Oregon or Columbia) as at Nootka Sound or any other point of that coast; & that trade be consequently conducted through the Missouri & U.S. more beneficially than by the circumnavigation now practised.

On your arrival on that coast endeavor to learn if there be any port within your reach frequented by the sea-vessels of any nation, and to send two or your trusted people back by sea, in such way as shall appear practicable, with a copy of your notes. And should you be of opinion that the return of your party by the way they went will be eminently dangerous, then ship the whole, & return by sea by way of Cape Horn or the Cape of Good Hope, as you shall be able. As you will be without money, clothes or provisions, you must endeavor to use the credit of the U.S. to obtain them; for which purpose open letters of credit shall be furnished you authorizing you to draw on the Executive of the U.S. or any of its officers in any part of the world, in which drafts can be disposed of, and to apply with our recommendations to the consuls, agents, merchants or citizens of any nation with which we have intercourse, assuring them in our name that any aids they may furnish you, shall be honorably repaid and on demand. Our consuls Thomas Howes at Batavia in Java, William Buchanan of the Isles of France and Bourbon & John Elmslie at the Cape of Good Hope will be able to supply your necessities by drafts on us.

Should you find it safe to return by the way you go, after sending two of your party round by sea, or with your whole party, if no conveyance by sea can be found, do so; making such observations on your return as may serve to supply, correct or confirm those made on your outward journey.

In re-entering the U.S. and reaching a place of safety, discharge any of your attendants who may desire & deserve it: procuring for them immediate paiment of all arrears of pay & cloathing which may have incurred since their departure & assure them that they shall be recommended to the liberality of the Legislature for the grant of a souldier's portion of land each, as proposed in my message to Congress: & repair yourself with your papers to the seat of government.

To provide, on the accident of your death, against anarchy, dispersion & the consequent danger to your party, and total failure of the enterprise, you are hereby authorized by any instrument signed & written in your own hand to name the person among them who shall succeed to the command on your decease, & by like instruments to change the nomination from time to time, as further experience of the characters accompanying you shall point out superior fitness: and all the powers & authorities given to yourself are, in the event of your death transferred to & vested in the successor so named, with further power to him, & his successors in like manner to name each his successor, who, on the death of his predecessor shall be invested with all the powers & authorities given to yourself.

Given under my hand at the city of Washington, this 20th day of June, 1803.

QUESTIONS AND SUGGESTIONS

1. Summarize Jefferson's directions to Lewis in the style of an executive summary.

2. Prepare the necessary visual aids to accompany this text for oral presentation.

3. Write a longer essay on any of the following topics:
    a. Methods of navigating and exploring a river
    b. The history and use of the chronometer
    c. The history and use of the orrery
    d. Thomas Jefferson as a scientist

For each of these topics, you may wish to consult David S. Landes, *Revolution in Time: Clocks and the Making of the Modern World* (Harvard, 1983).

ALMIRA PHELPS

Galvanism

ALMIRA PHELPS (1793–1884) began teaching at the age of 16, after being educated, largely at home, in classical languages, mathematics, and sciences. Married and widowed twice, she was first a teacher then a principal of various women's academies. Phelps retired from education in 1856 and is best known for her progressive ideas on female education and for success in reforming it from "polite learning" to more substantial education for girls in science, mathematics, and the classics. The second woman to gain membership in the American Association for the Advancement of Science, she was also a member of the Maryland Academy of Science. Among her works are *Familiar Lectures on Botany* (1829); *Dictionary of Chemistry* (1830); *Geology for Beginners* (1836); and *Lectures on Chemistry* (1837), from which the following chapter on Galvanism is excerpted. (Figures, paragraphs, and notes have been renumbered.)

Galvanism, or Voltaic Electricity

1. The subject of *Electricity* is intimately connected both with Natural Philosophy and Chemistry. As a *chemical* agent, it is chiefly confined to the department of *Galvanism*.

History of Galvanism. The earliest notice of any fact connected with Galvanism is found in a book entitled "The general theory of Pleasures," published in 1767, by a German metaphysician named Sultzer. It is there mentioned that a peculiar taste is excited when two slips of different metals, one lying on the tongue, and the other under it, are made to touch each other. This writer, however, gave no satisfactory explanation of the phenomena; and it attracted little attention till 1790, when it acquired importance from the discovery made by Galvani, a distinguished professor of Anatomy at Bologna; this philosopher had for some time, entertained the opinion that electricity was concerned in producing the muscular motions of animals; and his belief was strengthened, by observing that when the limbs of some recently skinned frogs, lying

---

1 Electricity as a chemical agent. History of Galvanism. Observations and experiments of Galvani.

190

on his table, were accidentally touched with a knife (the electric machine being in operation at the same time) convulsive motions were produced. In pursuing his researches on the subject, he found that the same result would be obtained, whenever two metals were made to touch each other, while one was in contact with the *nerves,* and the other metal in contact with the *muscles* of the frog.

Galvani concluded that the nerves and muscles of living animals are charged with electricity developed in the brain; and that, whenever a communication is made between them by means of a conductor, the equilibrium is restored, and motion produced.

2. Among the many who zealously examined and discussed these new phenomena, was *Volta,* a distinguished electrician of Pavia. He made numerous experiments on the subject, and arrived at the conclusion that the motions were, indeed, produced by electricity, not generated, however, as Galvani supposed, in the animal system, *but excited by the contact of the metals,* themselves. But notwithstanding the identity, now well established, of common electricity with the *Animal electricity* of Galvani, certain modifications of its character and applications, as well as the different modes of producing it, have caused the name *Galvanism* to be still retained. Volta constructed the *pile* of metallic plates which is distinguished by his name, and has greatly contributed to the advancement of chemical science; it is now superseded by improvements upon the original invention.

> Before we describe the effects of the Voltaic pile, it is necessary to premise the following facts.
>
> 1st. Whenever two plates of *different metals,* are made to touch each other by their broad surfaces, electric excitement is produced.
>
> 2nd. If the plates be separated by means of an insulating handle, one of them will be found *positively* and the other *negatively* excited.
>
> 3d. The plates being of equal surfaces, the positive excitement of one will be equal in intensity to the negative excitement of the other.
>
> 4th. Other circumstances being the same, the degree of

2 Conclusions of Volta. Voltaic pile. Facts connected with the development of electricity by means of the Voltaic pile, or Galvanism.

excitement will be greater, as the metals differ more in their degrees of affinity for oxygen.

5th. The more oxydable of the two metals will always become *positively* excited, and the other *negatively*.

3. Zinc and copper are the metals commonly used in galvanic experiments.

The cut represents a vessel (Fig. 1) containing an acid, much diluted with water, and two plates, the one of zinc, the other of copper (as shown by the letters Z and C); to each plate is soldered a piece of wire, the two ends of which meet in the center, opposite the place of insertion. This is called a *simple galvanic circle*. It is supposed that when in operation, there is, in such a circle, a continued current of electricity, flowing in the direction of the *arrows* from the zinc to the fluid, from the fluid to the copper, and from the copper back to the zinc. When the wires do not communicate, the galvanic circle is said to be broken. The wire attached to the copper plate is *negative,* that to the zinc plate *positive.* The wires are also called *poles;* thus we say, the *copper is the negative pole,* and the *zinc, the positive pole.*

4. The quantity of electricity developed by a single pair of plates being very small, Volta endeavored to devise means of increasing

Figure 1.

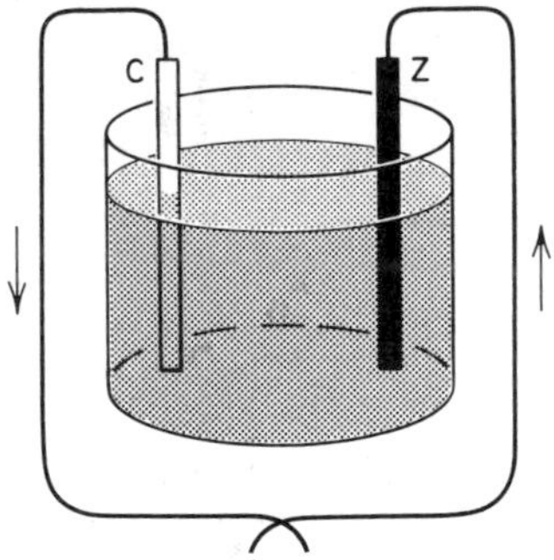

3 Metals commonly used in galvanic experiments. Galvanic circle. Negative and positive poles.
4 Construction of the Voltaic pile. Number and arrangement of plates.

it; and was finally led to the invention of the *Voltaic pile*. This instrument consists of *pairs* of zinc and copper plates, one above another. Each pair is called a *simple element* of the pile, the whole consisting of a *series* of simple elements or circles. Between each pair of plates is placed a piece of cloth wet with weak acid.

Figure 2 represents a *Voltaic pile,* commencing with zinc, Z, and ending with copper, C. The wires which meet in the center are the two poles. The direction in which the arrows point is that of the electric current. In constructing the Voltaic pile, from thirty to fifty plates of copper, and as many of zinc, are generally used; these are placed in regular order, each pair of plates being separated by a piece of cloth; thus a regular succession of copper, zinc, and cloth, is kept up through the whole series. The pieces of cloth should be somewhat smaller than the metallic plates, and should not be so moist as to yield any of the liquid by the pressure of the superincumbent metals. The pile is contained in a frame composed of a base and cap of dry varnished wood, connected by stout rods of glass.

5. The *pile* is capable of affecting the electrometer, and of producing muscular action in a much greater degre than the *single pair of plates*. The zinc being positive and copper negative, the electric equilibrium is restored on bringing them into communication by means of wires connected with each; as it is when the coatings of a charged Leyden phial are connected. But the causes of excitement being still in the pile, the equilibrium is instantly disturbed again, so that a continuous stream of the galvanic fluid is

Figure 2.

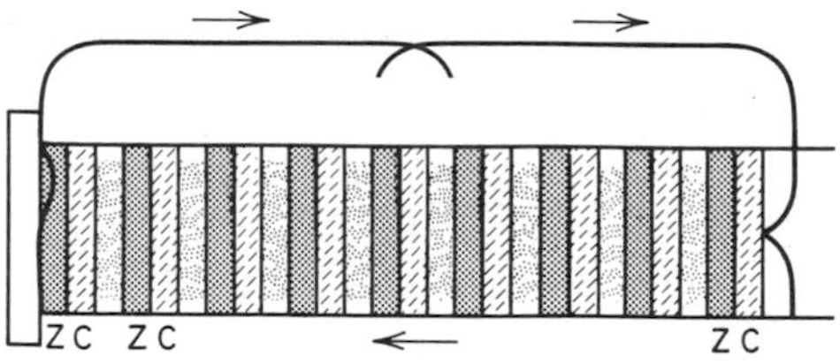

5 Action of the Voltaic pile. Differences and resemblances in the action of the Voltaic pile and the electrical machine.

produced: that is one of the most striking differences between the action of the pile and that of the common electric machine. By means of the pile, the Leyden phial may be charged, and the effects of the latter are precisely the same as when charged in the usual manner. If the two extremes of the pile are touched at once by the moistened fingers, a shock is felt differing little in kind from that produced by the phial, but much less in degree; but if, after the first effect, the contact be still kept up, a continuous and painful *thrilling* sensation is perceived; and if the electric current traverses any wound, burn, or excoriation, it causes it to smart severely. Volta remarked that the pain was greater on the side toward the negative pole; a circumstance in which Galvanism resembles common electricity.

6. Any number of piles may be combined by connecting the extreme copper plate, or *negative pole* of the first, with the extreme zinc plate, or *positive pole* of the second, and so on; and all the effects of the pile will be increased according to the number of *simple elements,* or pairs of plates.

This form of the pile is not now in use, as others more powerful and more convenient have been substituted.

7. One of the first of these was Cruickshank's trough (Fig. 3), commonly called the *Galvanic* battery; it consists of a trough of dry wood, divided into cells, by partitions placed at equal distances; each partition is made of a plate of zinc, and a plate of copper previously soldered together and fitted accurately into a groove cut in the wood; the joints being made perfectly tight with cement. The same order of arrangement must be observed as in the pile. This instrument is put into operation by filling all the cells about two thirds full of a saline or acid solution. Its effects are more powerful than those of the pile and may be increased by connecting several troughs in the manner we have just described (see § 6).

8. The actual contact of the metals is not necessary, as is seen in the construction of Volta's chain of cups (Fig. 4), commonly known as the *"Couronne des Tasses."** This arrangement, of which

6 Connection of piles.
7 Galvanic battery.
8 Couronne des Tasses.
* *Couronne des Tasses,* pronounced *kouron da tas,* literally a *crown of cups.*

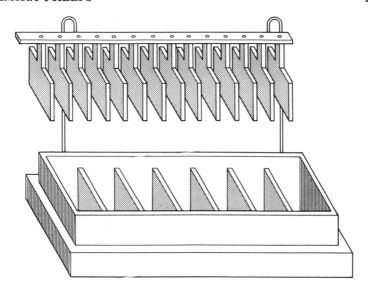

Figure 3.

the effects are greater than those of a pile of equal dimensions, consists of any number of pairs of zinc and copper plates generally about one and a half inch square; the zinc and copper plate of each pair are connected by a wire bent in the form of an inverted U: and the different elements are immersed in cups or glasses of acid, or saline liquids, in such a manner that the zinc plate of one pair shall be in the same vessel with the copper of the succeeding. Thus

Figure 4.

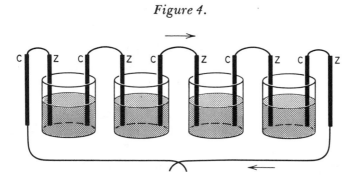

the different cups are connected only by the wire which joins the two members of each element: and the different elements act on each other only through the medium of the intervening liquid. The two extreme plates which are not immersed in the liquid, are not taken into the account, and may be used as means of connecting the two poles of the row.

The electricity is supposed to be excited by the mutual action of the surfaces of zinc and copper, opposed to each other in the same cup, and is conveyed from one cup to another by means of the wire which connects two successive cups.

9. *Modifications of the Galvanic Battery.* Dr. Wollaston observing that in the trough of Cruickshank, the effect of one zinc and one copper surface in each pair was lost, by soldering them together, he proposed to use double copper plates or to bend the copper plate so that it should entirely surround the zinc one, but without allowing the two to touch each other. In this way each zinc surface z, is opposed to one of copper, c, and the power is increased by one half. Batteries are now generally constructed on this principle: and a further improvement is made by connecting all the plates to a bar of dry wood, by means of which they can be removed at pleasure; the trough is made of porcelain or some other nonconductor, and is divided into cells by partitions of the same material.

The largest battery ever constructed is that of Mr. Children, the plates of which were 6 ft. long, and 2 ft. wide.

Hare's *Calorimeter,* or mover of heat (Fig. 5A), consists of a number of square zinc and copper plates of any convenient size, alternating with each other in a wooden frame. Two rectangular tubs accompany the apparatus, one containing the liquid acid for exciting the electricity, and the other to hold water for washing the plates. By means of an upright crossbar with a rope and pulley, the frame containing the plates may be at pleasure immersed in, or removed from the acid liquid: and this facility affords great advantages: for it is ascertained that the greatest action of the galvanic battery is at the first instant of immersion; and it is, therefore, important to be able to immerse all the plates at once. Besides the effects of the calorimeter in igniting wires, during its

9 [. . .] improved battery. Mr. Children's battery. Calo.

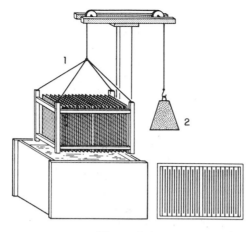

Figure 5A.

action much hydrogen is evolved from the liquid and the gas is sometimes inflamed by the great heat produced. (Figure 5A, at 1, represents the entire instrument ready to be plunged, and at 2, the top
of the plates.)

Another modification of the battery by Dr. Hare is called the
Deflagrator from its great power of burning the metals.

The battery most used at present, and which seems likely to su-
persede the use of all others, is called Groves' Constant Battery, a
section of which is represented at Figure 5B, a, a, a, the outer glass
vessels or tumblers; b, b, b, cylinders of zinc, open above and be-

Figure 5B.

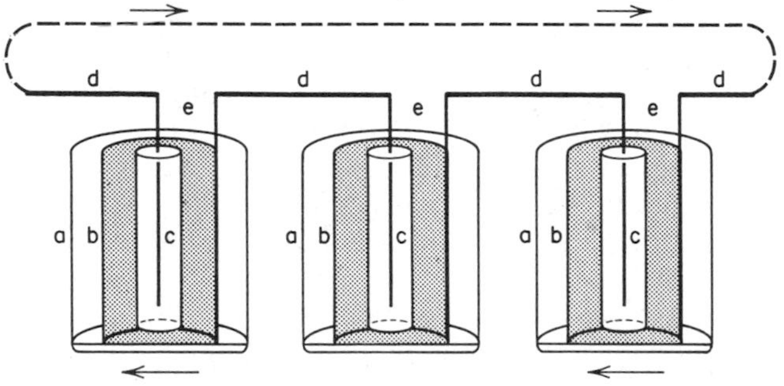

low; c, c, c, porcelain cylinders closed at the bottom, to receive the platina slips; d, d, d, bars of zinc connecting the zinc cylinder in one tumbler with the platina slip of the next; e, e, e, platina slips attached to the end of the zinc bars.

When in use, the outer glass vessel is filled with dilute sulphuric acid, and the inner porcelain vessel with strong nitric acid, and a connection being established between the platina slip at one extremity and the zinc cylinder at the other as represented by the dotted lines, the galvanic current then flows in the direction indicated by the arrows.

10. *Effects of Galvanism.* While the phenomena of galvanism and electricity seem to be produced by the *same agent,* they differ remarkably in the following particulars, viz: 1st, in the greater quantity of the electric fluid developed by the galvanic battery, 2nd, in its low intensity, and 3d, in the incessant renewal of the excitement as often as the equilibrium is restored, so as to produce a continuous current. To the last circumstance is attributable the superiority of galvanism over electricity in producing chemical decomposition.

11. The *igniting effects* of the galvanic pile are very remarkable.

When the two poles of a battery in action are connected by a small wire, the latter becomes intensely heated and gives out a light so vivid, that the eye can scarcely endure it. With a powerful battery, substances not fusible by any other means, are melted almost instantly. Even platinum is melted by it, as easily as wax by the flame of a candle. Of all substances, charcoal emits the most intensely, brilliant light.

> *Exp.* Two slender slips of dense charcoal, or of plumbago,* should be selected, scraped to a point, and fixed to each of the connecting wires. The battery being now put into action, and the charcoal points made to touch by means of insulating handles attached to the connecting wires, they immediately become vividly ignited; and if very slowly separated, an arc of intense light will fill the space between them. With the great battery of the Royal Institution at London, consisting of 2000 pairs of plates an arc appeared

10 Difference between electricity and galvanism.
11 Igniting effects of galvanism. *Exp.*
* Black lead.

four inches in length, and the heat existing there was so great as to fuse whatever substance was placed in it. Wires, even of the least oxidable of the metals, being made the medium of connection between the poles, may be burnt almost instantly.

12. Brilliant combustions may be exhibited in the following manner. Place some mercury in a flat dish and connect it with the negative pole of a strong battery; attach to the positive wire whatever is to be subjected to experiment, and make it touch the surface of the mercury. A point of fine iron wire burns in this manner with great rapidity, giving off vivid sparks in all directions, and producing an appearance like that of a brilliant star; the reflection from the bright surface of the mercury adding to its beauty. Gold-leaf burns with a beautiful green light the instant it touches the mercury, and is immediately converted into the purple *oxide of gold.* Silver and copper leaf, and even platinum wire, undergo vivid combustion. It is necessary to keep the surface of the mercury quite clean during these experiments, that its conducting power may not be impaired.

13. If the two connecting wires of a battery are immersed in water, so that their points shall be within a short distance of each other, an effervescence, arising from the evolution of gas, will be immediately observed; and the experiment may be so conducted that the gas can be collected. When collected, it will be found to consist of oxygen and hydrogen, mixed in precisely the proportion for forming water. *The hydrogen is always evolved at the negative pole,* and *the oxygen at the positive,* and by a contrivance of Dr. Wollaston, they may be collected and exhibited separately.

14. In procceding to consider the decomposing effects of galvanism, it is necessary to explain some of the Chemical properties of acids and alkalies.

Chemical properties of acids and alkalies. Acids have the property of changing to *red,* the *blue* color of certain vegetable infusions, as that of violets, or of purple cabbage. *Alkalies,* on the contrary, change the same blue infusions *green;* and a color which

12 Combustion of substances placed on mercury by means of the battery.
13 Decomposition of water by means of the battery. Experiment with a bent glass tube. Experiment with two strait tubes.
14 Characteristics of acids, alkalies, salts.

has been changed by one of these substances may be restored by
a sufficient quantity of the other. *Salts* are chemical compounds
of an acid and an alkali; and when the two are united in proper
proportions, their characteristic properties are entirely disguised,
and the salt is called *neutral.*

15. If we dissolve in water, Glauber's salt or *sulphate of soda*
(composed of *sulphuric acid* and *soda*), and add to the solution
a little of the blue infusion of cabbage, the color will remain unaltered shewing that the compound is *neutral.* Let this solution
be subjected to the action of a galvanic battery of sufficient power,
the liquid at the positive tube, will soon become red, proving the
presence of an acid there, while at the same time, the negative
tube will exhibit an alkali, as will be shown by the liquid in it
becoming green. Now this acid and alkali could only arise from
the decomposition of the sulphate of soda; and we are able, otherwise, to show that the contents of the two tubes, being mixed, will
reproduce this salt. The sulphuric acid and soda are both compounds, each containing oxygen, united to a peculiar combustible
body;* and by a galvanic arrangement of high power, the acid and
alkali are resolved into their elementary parts, that is, to their
ultimate analysis.

16. In all cases of proximate† analysis of salts, by the galvanic
battery, the acid will be found at the *positive* pole, and the alkali
at the *negative;* and, whenever by ultimate analysis in this manner
we resolve a substance into elements, of which one is *combustible,*
and the other *non-combustible,* the *latter* will be found at the
positive, and the former at the *negative* pole.

Mr. Faraday, who has paid much attention to the subject
of galvanism, advances the following opinions:—

1st. That the poles have no attractive, or repulsive ten-

---

15 Decomposition of a salt by the galvanic battery. What would be the ultimate
analysis of Glauber's salt?

* Sulphuric acid is composed of *sulphur* and *oxygen,* soda of a metal called
sodium united with oxygen.

† The analysis of sulphate of soda into the acid and the alkali is the *proximate*
analysis; the farther analysis of sulphuric acid and soda into three simple elements is the *ultimate* or last analysis.

dency. He prefers the term *electrodos,*‡ which signifies the way or door for electric currents.

2d. When a compound is decomposed by galvanism it is said to be *electrolyzed;* || the substances capable of decomposition are called *electrolytes;* the elements of an electrolyte are called *ions.*§ Electro-negative substances as oxygen, chlorine, acids, &c., he calls *anions;*¶ the electro-positive as hydrogen, alkalies, metals, &c., he calls *calions.***

3d. Most of the simple elements are *ions,* that is capable of forming compounds decomposable by galvanism.

4th. A single *ion* by itself has no tendency to pass to either of the *electrodes,* or is indifferent to the voltaic currents.

5th. To account for the decomposition of water by galvanism, Mr. Faraday supposes a line of particles between the two *electrodes,* along which the current passes. When a particle of oxygen is evolved at the positive electrode, its hydrogen is not at once transferred to the opposite electrode, but unites with the oxygen of the contiguous particle of water, on the side towards which the positive current is moving, then it passes to the next, and so on till it arrives at the pole. A similar row of particles of oxygen start from the negative electrode at the same moment and combine successively with the particles of hydrogen as they pass them on their way to the positive pole or electrode. This process is supposed similar to what takes place in all cases of galvanic decomposition.

17. *Discoveries of Sir H. Davy.* The Galvanic battery became, in the skillful hands of Davy, the means of effecting the most brilliant discoveries. With this instrument, he ascertained the *com-*

‡ From the Greek *electron,* and *odos,* a way.
|| From *electron,* and *lus,* to unloose or disengage.
§ Pronounced *i-ons,* from *ion,* going, participle of the verb to go.
¶ From *ana,* upwards, and *odos,* the way in which the sun rises.
** From *kala,* downwards, the way in which the sun sets.
16 What takes place in all cases of proximate and ultimate analysis of salts by the galvanic battery? Mr. Faraday's theory of decomposition.
17 Decompositions effected by Davy. Strong proof of the elementary nature of a body.

pound nature of the alkalies and earths, a fact which, previously had been only suspected. This discovery has introduced a new era in the annals of Chemistry, and added several new metals to the list of simple elements. No known compound has been able to resist the decomposing power of galvanism; and it is regarded as the strongest proof of the elementary or simple nature of a body, when it gives no signs of decomposition on being subjected to the influence of this agent.

18. To account for the decomposing effect of galvanism, it is necessary to recur to the principle of electric attraction and repulsion. If two particles, united to form a compound molecule, are both brought into the *same electrical state,* they will exert a mutual repulsion; and if this repulsion be more powerful than the force of their chemical attraction, they must necessarily separate, and the compound will be destroyed. Or, even if they be in *opposite states of electricity,* since they are both within the influence of the battery, that particle which has the strongest tendency to become negatively electric, will naturally become so by induction, and be attracted to the positive pole; and at the same time, a similar change in the opposite direction will take place with the other particle.

19. Davy was led to infer that chemical and electrical attractions are effects of the same cause. Having brought a dry acid in contact with a metal, he found that the former became *electro-negative;* an alkali, treated in the same manner, became *electro-positive;* and when an acid and an alkali, both dry, were made to touch each other, electrical excitement was produced, the acid being negative and the alkali positive.

Furthermore, those bodies which exhibit the greatest tendency to chemical combination, are also prone to assume opposite electrical states; and though two bodies, *A* and *B,* which, if successively

---

18 Explanation of the decomposing effect of galvanism. Why compounds are destroyed by electrical repulsion. Why there should be a decomposition when the particles are in opposite states of electricity.

19 Davy's experiments to prove the connection between chemical and electrical attractions. Electricity of acids and alkalies. A body may be electro-positive with one body, and electro-negative with another. Why chlorine and oxygen form compounds which are easily decomposed.

brought in contact with C, would each assume the same electrical state, may be in opposite electrical states when in contact with each other; yet if they combine, their compound $A B$, is held together by weak affinities and is easily decomposed. Examples of this kind may be found in the instances of chlorine and oxygen; each of these bodies is strongly electro-negative when in contact with hydrogen, and each forms with it, a well-defined compound. But chlorine, though electro-negative when compared with almost all bodies, is electro-positive with regard to oxygen, and may be made to combine with it; yet the compounds formed by chlorine and oxygen are decomposed with remarkable facility.

20. The electro-chemical theory, supposes the same electrical excitement to take place between *atoms* in contact with each other, as we have seen to be produced by masses; that the atoms remain in contact; that is to say, in *combination*, in consequence of the electrical attraction consequent on this excitement; and that their union ceases, whenever, by any cause, they are brought into the same electrical state, or when they are exposed to the action of any third body which is more highly excited than either; for, in the latter case, the highly excited body will attract the particle which is dissimilarly excited, and repel that which is similarly so; and this is what happens in the decomposition of a compound substance by the galvanic battery.

21. Following the same course of reasoning which led him to the discovery of the alkaline metals,* Davy made other useful applications of his theory. Although the metals, compared with oxygen, are *all* electro-positive, yet when compared with each other, as in the case of zinc and copper, they may have opposite natural *electric energies;* and from experiment, as well as from theory, it is shown that the *positive metal will have the strongest tendency to combine with oxygen.* Thus copper is rapidly corroded in acid or saline solutions; but in contact with zinc, iron, and some other

20 Electro-chemical theory founded upon the preceding facts.

21 Applications made by Davy of his theory, with respect to the electrical attration of the metals. Why copper is protected from rust or oxidation, by zinc or iron. Iron and steel protected by zinc.

* Sodium, potassium, &c. being metals found in alkalies, soda, potash &c. are termed alkaline metals.

metals, copper becomes electro-negative and remains bright, while the other metal is rapidly oxidized; this happens, also, in the galvanic battery.

Davy found that a slip of zinc or iron, would protect from rust 150 times its surface of copper, though constantly exposed to the action of salt water; and he proposed to apply this principle to the preservation of the copper sheathing of ships. The rusting of fine iron or steel instruments, may be effectually prevented by fixing a piece of zinc in their handles or elsewhere, so that it shall be always in contact with the blade.

QUESTIONS AND SUGGESTIONS

1. Compare Phelps's discussion of Galvanism and Electricity to the discussions of these subjects found in a modern science textbook for college students. What similarities and differences do you notice in the comparison, especially pertaining to the authors' understandings of "college-level" readers?

2. What role does history play in the composition of a modern science textbook?

3. Consult The Dictionary of Scientific Biography for reports (oral or written) on Galvani, Volta, or Davy.

4. Compose your own history of the workings of an object similar to the battery. Possible subjects are the light bulb, the transistor, the water pump, or the radio. Or develop a process analysis on one of the objects above intended for the non-specialist college-educated reader.

RICHARD P. FEYNMAN

Physics: 1920 to Today

In the 1940s, RICHARD P. FEYNMAN (1918–1988) developed quantum electro-dynamics, in which the behavior of electrons was worked out with greater precision than it had been previously. J. S. Schwinger and Shinichiro Tomonaga did similar work independently, and all three shared the 1965 Nobel Prize in physics. Feynman was renowned as a teacher; his *Lectures on Physics*, from which this selection comes, were originally delivered to students at California Institute of Technology and are widely used as a textbook.

It is a little difficult to begin at once with the present view, so we shall first see how things looked in about 1920 and then take a few things out of that picture. Before 1920, our world picture was something like this: The "stage" on which the universe goes is the three-dimensional *space* of geometry, as described by Euclid, and things change in a medium called *time*. The elements on the stage are *particles*, for example the atoms, which have some *properties*. First, the property of inertia: if a particle is moving it keeps on going in the same direction unless *forces* act upon it. The second element, then, is *forces*, which were then thought to be of two varieties: First, an enormously complicated, detailed kind of interaction force which held the various atoms in different combinations in a complicated way, which determined whether salt would dissolve faster or slower when we raise the temperature. The other force that was known was a long-range interaction—a smooth and quiet attraction—which varied inversely as the square of the distance, and was called *gravitation*. This law was known and was very simple. *Why* things remain in motion when they are moving, or *why* there is a law of gravitation was, of course, not known.

A description of nature is what we are concerned with here. From this point of view, then, a gas, and indeed *all* matter, is a myriad of moving particles. Thus many of the things we saw while standing at the seashore can immediately be connected. First the pressure: this comes from the collisions of the atoms with the walls or whatever; the drift of the atoms, if they are all moving in one direction on the average, is wind; the *random* internal motions are the *heat*. There are waves of excess density, where too many particles have collected, and so as they rush off they push up piles of particles farther out, and so on. This wave of excess density is *sound*. It is a tremendous achievement to be able to understand so much. . . .

What *kinds* of particles are there? There were considered to be 92 at that time: 92 different kinds of atoms were ultimately discovered. They had different names associated with their chemical properties.

The next part of the problem was, *what are the short-range forces?* Why does carbon attract one oxygen or perhaps two oxygens, but not three oxygens? What is the machinery of interaction between atoms? Is it gravitation? The answer is no. Gravity is entirely too weak. But imagine a force analogous to gravity, varying inversely with the square of the distance, but enormously more powerful and having one difference. In gravity everything attracts everything else, but now imagine that there are *two kinds* of "things," and that this new force (which is the electrical force, of course) has the property that likes *repel* but unlikes *attract*. The "thing" that carries this strong interaction is called *charge*.

Then what do we have? Suppose that we have two unlikes that attract each other, a plus and a minus, and that they stick very close together. Suppose we have another charge some distance away. Would it feel any attraction? It would feel *practically none,* because if the first two are equal in size, the attraction for the one and the repulsion for the other balance out. Therefore there is very little force at any appreciable distance. On the other hand, if we get *very close* with the extra charge, *attraction* arises, because the repulsion of likes and attraction of unlikes will tend to bring unlikes closer together and push likes farther apart. Then the repulsion will be *less* than the attraction. This is the reason why the atoms, which are constituted out of plus and minus electric charges, feel very little force when they are separated by appreciable distance (aside from gravity). When they come close together, they can "see inside" each other and rearrange their charges, with the result that they have a very strong interaction. The ultimate basis of an interaction between the atoms is *electrical.* Since this force is so enormous, all the plusses and all minuses will normally come together in as intimate a combination as they can. All things, even ourselves, are made of fine-grained, enormously strongly interacting plus and minus parts, all neatly balanced out. Once in a while, by accident, we may rub off a few minuses or a few plusses (usually it is easier to rub off minuses), and in those circumstances we find the force of electricity *unbalanced,* and we can then see the effects of these electrical attractions.

To give an idea of how much stronger electricity is than gravitation, consider two grains of sand, a millimeter across, thirty meters apart. If the force between them were not balanced, if everything attracted everything else instead of likes repelling, so that there were no cancellation, how much force would there be? There would be a force of *three million tons* between the two! You see, there is very, *very* little excess or deficit of the number of negative or positive charges necessary to produce appreciable electrical effects. This is, of course, the reason why you cannot see the difference between an electrically charged or uncharged thing—so few particles are involved that they hardly make a difference in the weight or size of an object.

With this picture the atoms were easier to understand. They were thought to have a "nucleus" at the center, which is positively electrically charged and very massive, and the nucleus is surrounded by a certain number of "electrons" which are very light and negatively charged. Now we go a little ahead in our story to remark that in the nucleus itself there were found two kinds of particles, protons and neutrons, almost of the same weight and very heavy. The protons are electrically charged and the neutrons are neutral. If we have an atom with six protons inside its nucleus, and this is surrounded by six electrons (the negative particles in the ordinary world of matter are all electrons, and these are very light compared with the protons and neutrons which make nuclei), this would be atom number six in the chemical table, and it is called carbon. Atom number eight is called oxygen, etc., because the chemical properties depend upon the electrons on the *outside,* and in fact only upon *how many* electrons there are. So the *chemical* properties of a substance depend only on a number, the num-

other parts from a physicist's standpoint, but from a human stand-point, of course, it *is* more interesting. If we go up even higher in frequency, we get x-rays. X-rays are nothing but very high-frequency light. If we go still higher, we get gamma rays. These two terms, x-rays and gamma rays, are used almost synonymously. Usually electromagnetic rays coming from nuclei are called gamma rays, while those of high energy from atoms are called x-rays, but at the same frequency they are indistinguishable physically, no matter what their source. If we go to still higher frequencies, say to 10^{24} cycles per second, we find that we can make those waves artificially, for example with the synchrotron here at Caltech. We can find electromagnetic waves with stupendously high fre-quencies—with even a thousand times more rapid oscillation—in the waves found in *cosmic rays*. These waves cannot be controlled by us.

Quantum Physics

Having described the idea of the electromagnetic field, and that this field can carry waves, we soon learn that these waves actually behave in a strange way which seems very unwavelike. At higher frequencies they behave much more like *particles!* It is *quantum mechanics,* discovered just after 1920, which explains this strange behavior. In the years before 1920, the picture of space as a three-

Table 1. The Electromagnetic Spectrum.

Frequency in oscillations/sec	Name	Rough behavior
10^2	Electrical disturbance	Field
$5 \times 10^5 - 10^6$	Radio broadcast	
10^8	FM–TV	
10^{10}	Radar	Waves
$5 \times 10^{14} - 10^{15}$	Light	
10^{18}	X-rays	
10^{21}	γ-rays, nuclear	
10^{24}	γ-rays, "artificial"	Particle
10^{27}	γ-rays, in cosmic rays	

dimensional space, and of time as a separate thing, was changed by Einstein, first into a combination which we call space-time, and then still further into a *curved* space-time to represent gravitation. So the "stage" is changed into space-time, and gravitation is presumably a modification of space-time. Then it was also found that the rules for the motions of particles were incorrect. The mechanical rules of "inertia" and "forces" are *wrong*—Newton's laws are *wrong*—in the world of atoms. Instead, it was discovered that things on a small scale behave *nothing like* things on a large scale. That is what makes physics difficult—and very interesting. It is hard because the way things behave on a small scale is so "unnatural"; we have no direct experience with it. Here things behave like nothing we know of, so that it is impossible to describe this behavior in any other than analytic ways. It is difficult, and takes a lot of imagination.

Quantum mechanics has many aspects. In the first place, the idea that a particle has a definite location and a definite speed is no longer allowed; that is wrong. To give an example of how wrong classical physics is, there is a rule in quantum mechanics that says that one cannot know both where something is and how fast it is moving. The uncertainty of the momentum and the uncertainty of the position are complementary, and the product of the two is constant. We can write the law like this: $\Delta x \, \Delta p \geq h/2\pi$, but we shall explain it in more detail later. This rule is the explanation of a very mysterious paradox: if the atoms are made out of plus and minus charges, why don't the minus charges simply sit on top of the plus charges (they attract each other) and get so close as to completely cancel them out? *Why are atoms so big?* Why is the nucleus at the center with the electrons around it? It was first thought that this was because the nucleus was so big; but no, the nucleus is *very small*. An atom has a diameter of about 10^{-8} cm. The nucleus has a diameter of about 10^{-13} cm. If we had an atom and wished to see the nucleus, we would have to magnify it until the whole atom was the size of a large room, and then the nucleus would be a bare speck which you could just about make out with the eye, but very nearly *all the weight* of the atom is in that infinitesimal *nucleus*. What keeps the electrons from simply falling in? This principle: If they were in the nucleus, we would know their position precisely, and the uncertainty principle would then require that they have a very *large* (but uncertain) momentum, i.e., a very large *kinetic energy*. With this energy they would break away from the nucleus. They make a compromise: they leave themselves a little room for this uncertainty and then jiggle with a certain amount of minimum motion in accordance with this rule. (Remember that when a crystal is cooled to absolute zero, . . . the atoms do not stop moving, they still jiggle. Why? If they stopped moving, we would know where they were and that they had zero motion, and that is against the uncertainty principle. We cannot know where they are and how fast they are moving, so they must be continually wiggling in there!)

Another most interesting change in the ideas and philosophy of science brought about by quantum mechanics is this: it is not possible to predict *exactly* what will happen in any circumstance. For example, it is possible to arrange an atom which is ready to emit light, and we can measure when it has emitted light by picking up a photon particle, which we shall describe shortly. We cannot, however, predict *when* it is going to emit the light or, with several atoms, *which one* is going to. You may say that this is because there are some internal "wheels" which we have not looked at closely enough. No, there *are* no internal wheels; nature, as we understand it today, behaves in such a way that it is *fundamentally impossible* to make a precise prediction of *exactly what will happen* in a given experiment. This is a horrible thing; in fact, philosophers have said before that one of the fundamental requisites of science is that whenever you set up the same conditions, the same thing must happen. This is simply *not true,* it is *not* a fundamental condition of science. The fact is that the same thing does not happen, that we can find only an average, statistically, as to what happens. Nevertheless, science has not completely collapsed. Philosophers, incidentally, say a great deal about what is *absolutely necessary* for science, and it is always, so far as one can see, rather naive, and probably wrong. For example, some philosopher or other said it is fundamental to the scientific effort that if an experiment is performed in, say, Stockholm, and then the same experiment is done in, say, Quito, the *same results* must occur. That is quite false. It is not necessary that *science* do that; it may be a *fact of experience,* but it is not necessary. For example, if one of the experiments is to look out at the sky and see the aurora borealis in Stockholm, you do not see it in Quito; that is a different phenomenon. "But," you say, "that is something that has to do with the outside; can you close yourself up in a box in Stockholm and pull down the shade and get any difference?" Surely. If we take a pendulum on a universal joint, and pull it out and let go, then the pendulum will swing almost in a plane, but not quite. Slowly the plane keeps changing in Stockholm, but not in Quito. The blinds are down, too. The fact that this happened does not bring on the destruction of science. What *is* the fundamental hypothesis of science, the fundamental philosophy? We stated it in the first chapter: *the sole test of the validity of any idea is experiment.* If it turns out that most experiments work out the same in Quito as they do in Stockholm, then those "most experiments" will be used to formulate some general law, and those experiments which do not come out the same we will say were a result of the environment near Stockholm. We will invent some way to summarize the results of the experiment, and we do not have to be told ahead of time what this way will look like. If we are told that the same experiment will always produce the same result, that is all very well, but if when we try it, it does *not*, then it does *not*. We just have to take what we see, and then formulate all the rest of our ideas in terms of our actual experience.

Returning again to quantum mechanics and fundamental physics, we cannot go into details of the quantum-mechanical principles at this time, of course, because these are rather difficult to understand. We shall assume that they are there, and go on to describe what some of the consequences are. One of the consequences is that things which we used to consider as waves also behave like particles, and particles behave like waves; in fact everything behaves the same way. There is no distinction between a wave and a particle. So quantum mechanics *unifies* the idea of the field and its waves, and the particles, all into one. Now it is true that when the frequency is low, the field aspect of the phenomenon is more evident, or more useful as an approximate description in terms of everyday experiences. But as the frequency increases, the particle aspects of the phenomenon become more evident with the equipment with which we usually make the measurements. In fact, although we mentioned many frequencies, no phenomenon directly involving a frequency has yet been detected above approximately 10^{12} cycles per second. We only *deduce* the higher frequencies from the energy of the particles, by a rule which assumes that the particle-wave idea of quantum mechanics is valid.

Thus we have a new view of electromagnetic interaction. We have a new kind of *particle* to add to the electron, the proton, and the neutron. That new particle is called a *photon*. The new view of the interaction of electrons and protons that is electromagnetic theory, but with everything quantum-mechanically correct, is called *quantum electrodynamics*. This fundamental theory of the interaction of light and matter, or electric field and charges, is our greatest success so far in physics. In this one theory we have the basic rules for all ordinary phenomena except for gravitation and nuclear processes. For example, out of quantum electrodynamics come all known electrical, mechanical, and chemical laws: the laws for the collision of billiard balls, the motions of wires in magnetic fields, the specific heat of carbon monoxide, the color of neon signs, the density of salt, and the reactions of hydrogen and oxygen to make water are all consequences of this one law. All these details can be worked out if the situation is simple enough for us to make an approximation, which is almost never, but often we can understand more or less what is happening. At the present time no exceptions are found to the quantum-electrodynamic laws outside the nucleus, and there we do not know whether there is an exception because we simply do not know what is going on in the nucleus.

In principle, then, quantum electrodynamics is the theory of all chemistry, and of life, if life is ultimately reduced to chemistry and therefore just to physics because chemistry is already reduced (the part of physics which is involved in chemistry being already known). Furthermore, the same quantum electrodynamics, this great thing, predicts a lot of new things. In the first place, it tells the properties of very high-energy photons, gamma rays, etc. It predicted another very remarkable thing: besides the electron, there should be another particle of the same mass, but of opposite charge, called a *positron,* and these two, coming together, could annihilate each other with the emission of light or gamma rays. (After all, light and gamma rays are all the same, they are just different points on a frequency scale.) The generalization of this, that for each particle there is an antiparticle, turns out to be true. In the case of electrons, the antiparticle has another name—it is called a positron, but for most other particles, it is called anti-so-and-so, like antiproton or antineutron. In quantum electrodynamics, *two numbers* are put in and most of the other numbers in the world are supposed to come out. The two numbers that are put in are called the mass of the electron and the charge of the electron. Actually, that is not quite true, for we have a whole set of numbers for chemistry which tells how heavy the nuclei are. That leads us to the next part.

Nuclei and Particles

What are the nuclei made of, and how are they held together? It is found that the nuclei are held together by enormous forces. When these are released, the energy released is tremendous compared with chemical energy, in the same ratio as the atomic bomb explosion is to a TNT explosion, because, of course, the atomic bomb has to do with changes inside the nucleus, while the explosion of TNT has to do with the changes of the electrons on the outside of the atoms. The question is, what are the forces which hold the protons and neutrons together in the nucleus? Just as the electrical interaction can be connected to a particle, a photon, Yukawa suggested that the forces between neutrons and protons also have a yield of some kind, and that when this field jiggles it behaves like a particle. Thus there could be some other particles in the world besides protons and neutrons, and he was able to deduce the properties of these particles from the already known characteristics of nuclear forces. For example, he predicted they should have a mass of two or three hundred times that of an electron; and lo and behold, in cosmic rays there was discovered a particle of the right mass! But it later turned out to be the wrong particle. It was called a μ-meson, or muon.

However, a little while later, in 1947 or 1948, another particle was found, the π-meson, or pion, which satisfied Yukawa's criterion. Besides the proton and the neutron, then, in order to get nuclear forces we must add the pion. Now, you say, "Oh great!, with this theory we make quantum nucleodynamics using the pions just like Yukawa wanted to do, and see if it works, and everything will be explained." Bad luck. It turns out that the calculations that are involved in this theory are so difficult that no one has ever been able to figure out what the consequences of the theory are, or to check it against experiment, and this has been going on now for almost twenty years!

So we are stuck with a theory, and we do not know whether it is right or wrong, but we do know that it is a *little* wrong, or at least incomplete. While we have been dawdling around theoretically, trying to calculate the consequences of this theory, the experimentalists have been discovering some things. For example, they had already discovered this μ-meson or muon, and we do not yet know where it fits. Also, in cosmic rays, a large number of other "extra" particles were found. It turns out that today we have approximately thirty particles, and it is very difficult to understand the relationships of all these particles, and what nature wants them for, or what the connections are from one to another. We do not today understand these various particles as different aspects of the same thing, and the fact that we have so many unconnected particles is a representation of the fact that we have so much unconnected information without a good theory. After the great successes of quantum electrodynamics, there is a certain amount of knowledge of nuclear physics which is rough knowledge, sort of half experience and half theory, assuming a type of force between protons and neutrons and seeing what will happen, but not really understanding where the force comes from. Aside from that, we have made very little progress. We have collected an enormous number of chemical elements. In the chemical case, there suddenly appeared a relationship among these elements which was unexpected, and which is embodied in the periodic table of Mendeléev. For example, sodium and potassium are about the same in their chemical properties and are found in the same column in the Mendeléev chart. We have been seeking a Mendeléev-type chart for the new particles. One such chart of the new particles was made independently by Gell-Mann in the U.S.A. and Nishijima in Japan. The basis of their classification is a new number, like the electric charge, which can be assigned to each particle, called its "strangeness," S. This number is conserved, like the electric charge, in reactions which take place by nuclear forces.

In table 2 are listed all the particles. We cannot discuss them

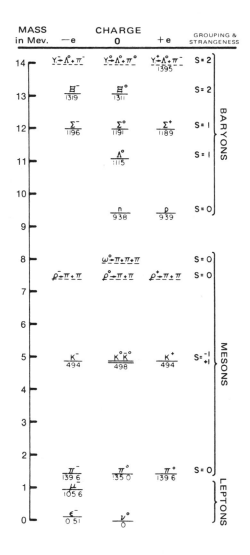

Table 2. Elementary Particles.

the field of gravity also has a quantum-mechanical analog (a quantum theory of gravitation has not yet been worked out), then there will be a particle, a graviton, which will have zero mass.

What is this "zero mass"? The masses given here are the masses of the particles *at rest.* The fact that a particle has zero mass means, in a way, that it cannot *be* at *rest.* A photon is never at rest, it is always moving at 186,000 miles a second. We will understand more what mass means when we understand the theory of relativity, which will come in due time.

Thus we are confronted with a large number of particles, which together seem to be the fundamental constituents of matter. Fortunately, these particles are not *all* different in their *interactions* with one another. In fact, there seem to be just *four kinds* of interaction between particles which, in the order of decreasing strength, are the nuclear force, electrical interactions, the beta-decay interaction, and gravity. The photon is coupled to all charged particles and the strength of the interaction is measured by some number, which is 1/137. The detailed law of this coupling is known, that is quantum electrodynamics. Gravity is coupled to all *energy,* but its coupling is extremely weak, much weaker than that of electricity. This law is also known. Then there are the so-called weak decays—beta decay, which causes the neutron to disintegrate into proton, electron, and neutrino, relatively slowly. This law is only partly known. The so-called strong interaction, the meson-baryon interaction, has a strength of 1 in this scale, and the law is completely unknown, although there are a number of known rules, such as that the number of baryons does not change in any reaction.

Table 3.   Elementary Interactions.

Coupling	Strength*	Law
Photon to charge particles	$\sim 10^{-2}$	Law known
Gravity to all energy	$\sim 10^{-40}$	Law known
Weak decays	$\sim 10^{-5}$	Law partly known
Mesons to baryons	~ 1	Law unknown (some rules known)

* The "strength" is a dimensionless measure of the coupling constant involved in each interaction ($\sim$ means "approximately").

This then, is the horrible condition of our physics today. To summarize it, I would say this: outside the nucleus, we seem to know all; inside it, quantum mechanics is valid—the principles of quantum mechanics have not been found to fail. The stage on which we put all of our knowledge, we would say, is relativistic space-time; perhaps gravity is involved in space-time. We do not know how the universe got started, and we have never made experiments which check our ideas of space and time accurately, below some tiny distance, so we only *know* that our ideas work above that distance. We should also add that the rules of the game are the quantum-mechanical principles, and those principles apply, so far as we can tell, to the new particles as well as to the old. The origin of the forces in nuclei leads us to new particles, but unfortunately they appear in great profusion and we lack a complete understanding of their interrelationship, although we already know that there are some very surprising relationships among them. We seem gradually to be groping toward an understanding of the world of sub-atomic particles, but we really do not know how far we have yet to go in this task.

QUESTIONS AND SUGGESTIONS

1. Feynman's style is both casual and rushed. What does he convey about his sense of himself and of science by his tone? Point to features of the essay to support your answer.

2. Contrast Feynman's way of getting concepts across with Lynn Margulis's.

3. Feynman and Michael Faraday both chose the lecture hall as a forum for presenting these subjects. How does this affect their style and content? Do their "voices" come through in writing? Do any of the other authors in this book have distinctive "voices"?

4. Why does Feynman believe physics is in horrible condition?

CHARLES DARWIN

Keeling Islands: Coral Formations

CHARLES DARWIN (1809–1882) was born into a distinguished British scientific family: his father was a well-to-do physician, and his grandfather, Erasmus Darwin, was a poet-physician and amateur naturalist. Darwin showed no early intellectual promise and disappointed his family by discontinuing the study of medicine because he was horrified at the sight of operations performed upon children without anesthesia. But he did become interested in the study of natural history, and, after graduating from Cambridge, accepted a post as naturalist on the H.M.S. *Beagle* on a scientific expedition to southern latitudes. (He was not the Captain's first choice for the post.)

Traveling down the coast of South America, Darwin noticed small variations in species. Most striking of all were the variations of those species scattered among the Galapagos Islands, which gave him the material upon which to work out his evolutionary ideas. After his return to England, he published his notebooks, *A Naturalist's Voyage on the Beagle* (1839). The theory of coral reefs advanced in that book and here reprinted formed a model upon which to develop his later evolutionary theory. Twenty years later, he published *On the Origin of Species* (1859), and a furor ensued over whether evolution applied to man. But Darwin stood his ground and, in 1871, published *The Descent of Man.*

"Keeling Islands: Coral Formations" was read before the Geological Society in May 1837, was published as a part of the 1839 journal, and later was published as a separate volume, *The Structure and Distribution of Coral Reefs.* It is a small jewel, and exemplifies Darwin's style of exposition.

I will now give a very brief account of the three great classes of coral-reefs; namely, Atolls, Barrier, and Fringing-reefs, and will explain my views on their formation. Almost every voyager who has crossed the Pacific has expressed his unbounded astonishment at the lagoon-islands, or as I shall for the future call them by their Indian name of atolls, and has attempted some explanation. Even as long ago as the year 1605, Pyrard de Laval well exclaimed, "C'est une merveille de voir chacun de ces atollons, environné d'un grand banc de pierre tout autour, n'y ayant point d'artifice humain." The sketch of Whitsunday Island in the Pacific, copied from Capt. Beechey's admirable *Voyage*, gives but a faint idea of a faint idea of the singular aspect of an atoll: it is one of the smallest size, and has its narrow islets united together in a ring. The immensity of the ocean, the fury of the breakers, contrasted with the lowness of the land and the smoothness of the bright green water within the lagoon, can hardly be imagined without having been seen.

The earlier voyagers fancied that the coral-building animals instinctively built up their great circles to afford themselves protection in the inner parts; but so far is this from the truth, that those massive kinds, to whose growth on the exposed outer shores the very existence of the reef depends, cannot live within the lagoon, where other delicately-branching kinds flourish. Moreover, on this view, many species of distinct genera and families are supposed to combine for one end; and of such a combination, not a single instance can be found in the whole of nature. The theory that has been most generally received is, that atolls are based on sub-marine craters; but when we consider the form and size of some, the number, proximity, and relative positions of others, this idea loses its plausible character: thus, Suadiva Atoll is 44 geographical miles in diameter in one line, by 34 miles in another line; Rimsky is 54 by 20 miles across, and it has a strangely sinuous margin; Bow Atoll is 30 miles long, and on an average only 6 in width; Menchicoff Atoll consists of three atolls united or tied together. This theory, moreover, is totally inapplicable to the northern Maldiva Atolls in the Indian Ocean (one of which is 88 miles in length, and between 10 and 20 in breadth), for they are not bounded like ordinary atolls by narrow reefs, but by a vast number of separate little atolls; other little atolls rising out of the great central lagoon-like spaces. A third and better theory was advanced by Chamisso, who thought that from the corals growing more vigorously where exposed to the open sea, as undoubtedly is the case, the outer edges would grow up from the general foundation before any other part, and that this would account for the ring or cup-shaped structure. But we shall immediately see, that in this, as well as in the crater-theory, a most important consideration has been overlooked, namely, on what have the reef-building corals, which cannot live at a great depth, based their massive structures?

Numerous soundings were carefully taken by Captain Fitz-Roy

on the steep outside of Keeling Atoll, and it was found that within ten fathoms, the prepared tallow at the bottom of the lead, invariably came up marked with the impressions of living corals, but as perfectly clean as if it had been dropped on a carpet of turf; as the depth increased, the impressions became less numerous, but the adhering particles of sand more and more numerous, until at last it was evident that the bottom consisted of a smooth sandy layer: to carry on the analogy of the turf, the blades of grass grew thinner and thinner, till at last the soil was so sterile, that nothing sprang from it. From these observations, confirmed by many others, it may be safely inferred that the utmost depth at which corals can construct reefs is between 20 and 30 fathoms. Now there are enormous areas in the Pacific and Indian Oceans, in which every single island is of coral formation, and is raised only to that height to which the waves can throw up fragments, and the winds pile up sand. Thus the Radack group of atolls is an irregular square, 520 miles long and 240 broad; the Low Archipelago is elliptic-formed, 840 miles in its longer, and 420 in its shorter axis: there are other small groups and single low islands between these two archipelagoes, making a linear space of ocean actually more than 4000 miles in length, in which not one single island rises above the specified height. Again, in the Indian Ocean there is a space of ocean 1500 miles in length, including three archipelagoes, in which every island is low and of coral formation. From the fact of the reef-building corals not living at great depths, it is absolutely certain that throughout these vast areas, wherever there is now an atoll, a foundation must have originally existed within a depth of from 20 to 30 fathoms from the surface. It is improbable in the highest degree that broad, lofty, isolated, steep-sided banks of sediment, arranged in groups and lines hundreds of leagues in length, could have been deposited in the central and profoundest parts of the Pacific and Indian Oceans, at an immense distance from any continent, and where the water is perfectly limpid. It is equally improbable that the elevatory forces should have uplifted throughout the above vast areas, innumerable great rocky banks within 20 to 30 fathoms, or 120 to 180 feet, of the surface of the sea, and not one single point above that level; for where on the whole face of the globe can we find a single chain of mountains, even a few hundred miles in length, with their many summits rising within a few feet of a given level, and not one pinnacle above it? If then the foundations, whence the atoll-building corals sprang, were not formed of sediment, and if they were not lifted up to the required level, they must of necessity have subsided into it; and this at once solves the difficulty. For as mountain after mountain, and island after island, slowly sank beneath the water, fresh bases would be successively afforded for the growth of the corals. It is impossible here to enter into all the necessary details, but I venture to defy any one to explain in any other manner, how it is possible that numerous islands should be distributed throughout vast areas—all the islands being low—all being built of corals, absolutely requiring a foundation within a limited depth from the surface.

Before explaining how atoll-formed reefs acquire their peculiar structure, we must turn to the second great class, namely, barrier-reefs. These either extend in straight lines in front of the shores of a continent or of a large island, or they encircle smaller islands; in both cases, being separated from the land by a broad and rather deep channel of water, analogous to the lagoon within an atoll. It is remarkable how little attention has been paid to encircling barrier-reefs; yet they are truly wonderful structures. The following sketch represents part of the barrier encircling the island of Bolabola in the Pacific, as seen from one of the central peaks. In this instance the whole line of reef has been converted into land; but usually a snow-white line of great breakers, with only here and there a single low islet crowned with cocoa-nut trees, divides the dark heaving waters of the ocean from the light-green expanse of the lagoon-channel. And the quiet waters of this channel generally bathe a fringe of low alluvial soil, loaded with the most beautiful productions of the tropics, and lying at the foot of the wild, abrupt, central mountains.

Encircling barrier-reefs are of all sizes, from three miles to no less than forty-four miles in diameter; and that which fronts one side, and encircles both ends, of New Caledonia, is 400 miles long. Each reef includes one, two, or several rocky islands of various heights; and in one instance, even as many as twelve separate islands. The reef runs at a greater or less distance from the included land; in the Society Archipelago generally from one to three or four miles; but at Hogoleu the reef is 20 miles on the southern

ledge round them under water, the present shores would have been invariably bounded by great precipices; but this is most rarely the case. Moreover, on this notion, it is not possible to explain why the corals should have sprung up, like a wall, from the extreme outer margin of the ledge, often leaving a broad space of water within, too deep for the growth of corals. The accumulation of a wide bank of sediment all round these islands, and generally widest where the included islands are smallest, is highly improbable, considering their exposed positions in the central and deepest parts of the ocean. In the case of the barrier-reef of New Caledonia, which extends for 150 miles beyond the northern point of the island, in the same straight line with which it fronts the west coast, it is hardly possible to believe, that a bank of sediment could thus have been straightly deposited in front of a lofty island, and so far beyond its termination in the open sea. Finally, if we look to other oceanic islands of about the same height and of similar geological constitution, but not encircled by coral-reefs, we may in vain search for so trifling a circumambient depth as 30 fathoms, except quite near to their shores; for usually land that rises abruptly out of water, as do most of the encircled and non-encircled oceanic islands, plunges abruptly under it. On what then, I repeat, are these barrier-reefs based? Why, with their wide and deep moat-like channels, do they stand so far from the included land? We shall soon see how easily these difficulties disappear.

We come now to our third class of Fringing-reefs, which will require a very short notice. Where the land slopes abruptly under water, these reefs are only a few yards in width, forming a mere ribbon or fringe round the shores: where the land slopes gently under the water the reef extends further, sometimes even as much as a mile from the land; but in such cases the soundings outside the reef always show that the submarine prolongation of the land is gently inclined. In fact the reefs extend only to that distance from the shore, at which a foundation within the requisite depth from 20 to 30 fathoms is found. As far as the actual reef is concerned, there is no essential difference between it and that forming a barrier or an atoll: it is, however, generally of less width, and consequently few islets have been formed on it. From the corals growing more vigorously on the outside, and from the noxious effect of the sediment washed inwards, the outer edge of the reef is the highest part, and between it and the land there is generally a shallow sandy channel a few feet in depth. Where banks of sediment have accumulated near to the surface, as in parts of the West Indies, they sometimes become fringed with corals, and hence in some degree resemble lagoon-islands or atolls; in the same manner as fringing-reefs, surrounding gently-sloping islands, in some degree resemble barrier-reefs.

No theory on the formation of coral-reefs can be considered satisfactory which does not include the three great classes. We have seen that we are driven to believe in the subsidence of these vast areas, interspersed with low islands, of which not one rises above the height to which the wind and waves can throw up matter, and yet are constructed by animals requiring a foundation, and that foundation to lie at no great depth. Let us then take an island surrounded by fringing-reefs, which offer no difficulty in their structure; and let this island with its reef, represented by the unbroken lines in the woodcut, slowly subside. Now as the island sinks down, either a few feet at a time or quite insensibly, we may safely infer, from what is known of the conditions favourable to the growth of coral, that the living masses, bathed by the surf on the margin of the reef, will soon regain the surface. The water, however, will encroach little by little on the shore, the island becoming lower and smaller, and the space between the inner edge of the reef and the beach proportionally broader. A section of the reef and island in this state, after a subsidence of several hundred

Figure 3.   AA. Outer edges of the fringing-reef, at the level of the sea. BB. The shores of the fringed island. A'A'. Outer edges of the reef, after its upward growth during a period of subsidence, now converted into a barrier, with islets on it. B'B'. The shores of the now encircled island. CC. Lagoon-channel.

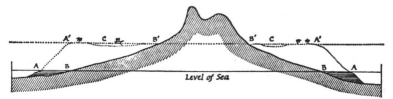

feet, is given by the dotted lines. Coral islets are supposed to have been formed on the reef; and a ship is anchored in the lagoon-channel. This channel will be more or less deep, according to the rate of subsidence, to the amount of sediment accumulated in it, and to the growth of the delicately branched corals which can live there. The section in this state resembles in every respect one drawn through an encircled island: in fact, it is a real section (on the scale of .517 of an inch to a mile) through Bolabola in the Pacific. We can now at once see why encircling barrier-reefs stand so far from the shores which they front. We can also perceive, that a line drawn perpendicularly down from the outer edge of the new reef, to the foundation of solid rock beneath the old fringing-reef, will exceed by as many feet as there have been feet of subsidence, that small limit of depth at which the effective corals can live: the little architects having built up their great wall-like mass, as the whole sank down, upon a basis formed of other corals and their consolidated fragments. Thus the difficulty on this head, which appeared so great, disappears.

If, instead of an island, we had taken the shore of a continent fringed with reefs, and had imagined it to have subsided, a great straight barrier, like that of Australia or New Caledonia, separated from the land by a wide and deep channel, would evidently have been the result.

Let us take our new encircling barrier-reef, of which the section is now represented by unbroken lines, and which, as I have said, is a real section through Bolabola, and let it go on subsiding. As the barrier-reef slowly sinks down, the corals will go on vigorously growing upwards; but as the island sinks, the water will gain inch by inch on the shore—the separate mountains first forming separate islands within one great reef—and finally, the last and highest pinnacle disappearing. The instant this takes place, a perfect atoll is formed: I have said, remove the high land from within an encircling barrier-reef, and an atoll is left, and the land has been removed. We can now perceive how it comes that atolls, having sprung from encircling barrier-reefs, resemble them in general size, form, in the manner in which they are grouped together, and in their arrangement in single or double lines; for they may be called rude outline charts of the sunken islands over which they stand. We can further see how it arises that the atolls in the Pacific and Indian oceans extend in lines parallel to the generally prevailing strike of the high islands and the great coast-lines of those oceans. I venture, therefore, to affirm, that on the theory of the upward growth of the corals during the sinking of the land, all the leading features in those wonderful structures, the lagoon-islands or atolls, which have so long excited the attention of voyagers, as well as in the no less wonderful barrier-reefs, whether encircling small islands or stretching for hundreds of miles along the shores of a continent, are simply explained.

It may be asked, whether I can offer any direct evidence of the subsidence of barrier-reefs or atolls; but it must be borne in mind how difficult it must ever be to detect a movement, the tendency of which is to hide under water the part affected. Nevertheless, at Keeling Atoll I observed on all sides of the lagoon old cocoa-nut trees undermined and falling; and in one place the foundation-posts of a shed, which the inhabitants asserted had stood seven years before just above high-water mark, but now was daily washed by every tide: on inquiry I found that three earthquakes, one of them severe, had been felt here during the last ten years. At Vanikoro, the lagoon-channel is remarkably deep, scarcely any alluvial soil has accumulated at the foot of the lofty included mountains, and remarkably few islets have been formed by the heaping of fragments and sand on the wall-like barrier-reef; these facts, and some analogous ones, led me to believe that this island must lately have subsided and the reef grown upwards: here again earthquakes are frequent and very severe. In the Society Archipelago, on the other hand, where the lagoon-channels are almost choked up, where much low alluvial land has accumulated, and where in some cases long islets have been formed on the barrier-reefs—facts all showing that the islands have not very lately subsided—only feeble shocks are most rarely felt. In these coral formations, where the land and water seem struggling for mastery, it must be ever difficult to decide between the effects of a change in the set of the tides and of a slight subsidence: that many of these reefs and atolls are subject to changes of some kind is certain; on some atolls the islets appear to have increased greatly within a late period; on others they have been partially or wholly washed away. The inhabitants of parts of the Maldiva Archipelago know the date of the first formation of some islets; in other parts, the corals are now flourishing on water-washed reefs, where holes made for graves attest the former existence of inhabited land. It is difficult to believe in frequent changes in the tidal currents of an open ocean; whereas, we have in the earthquakes recorded by the natives on some atolls, and in the great fissures observed on other atolls, plain evidence of changes and disturbances in progress in the subterranean regions.

It is evident, on our theory, that coasts merely fringed by reefs cannot have subsided to any perceptible amount; and therefore they must, since the growth of their corals, either have remained stationary or have been upheaved. Now it is remarkable how generally it can be shown, by the presence of upraised organic remains, that the fringed islands have been elevated: and so far, this is indirect evidence in favour of our theory. I was particularly struck with this fact, when I found to my surprise, that the descriptions given by MM. Quoy and Gaimard were applicable, not to reefs in general as implied by them, but only to those of the fringing-class; my surprise, however, ceased when I afterwards found that, by a strange chance, all the several islands visited by these eminent naturalists, could be shown by their own statements to have been elevated within a recent geological era.

Not only the grand features in the structure of barrier-reefs and of atolls, and of their likeness to each other in form, size, and other characters, are explained on the theory of subsidence—which theory we are independently forced to admit in the very areas in question, from the necessity of finding bases for the corals within the requisite depth—but many details in structure and exceptional cases can thus also be simply explained. I will give only a few instances. In barrier-reefs it has long been remarked with surprise, that the passages through the reef exactly face valleys in the included land, even in cases where the reef is separated from the land by a lagoon-channel so wide and so much deeper than the actual passage itself, that it seems hardly possible that the very small quantity of water or sediment brought down could injure the corals on the reef. Now, every reef of the fringing-class is breached by a narrow gateway in front of the smallest rivulet, even if dry during the greater part of the year, for the mud, sand, or gravel, occasionally washed down, kills the corals on which it is deposited. Consequently, when an island thus fringed subsides, though most of the narrow gateways will probably become closed by the outward and upward growth of the corals, yet any that are not closed (and some must always be kept open by the sediment and impure water flowing out of the lagoon-channel) will still continue to front exactly the upper parts of those valleys, at the mouths of which the original basal fringing-reef was breached.

We can easily see how an island fronted only on one side, or on one side with one end or both ends encircled by barrier-reefs, might after long-continued subsidence be converted either into a single wall-like reef, or into an atoll with a great straight spur projecting from it, or into two or three atolls tied together by straight reefs—all of which exceptional cases actually occur. As the reef-building corals require food, are preyed upon by other animals, are killed by sediment, cannot adhere to a loose bottom, and may be easily carried down to a depth whence they cannot spring up again, we need feel no surprise at the reefs both of atolls and barriers becoming in parts imperfect. The great barrier of New Caledonia is thus imperfect and broken in many parts; hence, after long subsidence, this great reef would not produce one great atoll 400 miles in length, but a chain or archipelago of atolls, of very nearly the same dimensions with those in the Maldiva Archipelago. Moreover, in an atoll once breached on opposite sides, from the likelihood of the oceanic and tidal currents passing straight through the breaches, it is extremely improbable that the corals, especially during continued subsidence, would ever be able again to unite the rim; if they did not, as the whole sank downwards, one atoll would be divided into two or more. In the Maldiva Archipelago there are distinct atolls so related to each other in position, and separated by channels either unfathomable or very deep (the channel between Ross and Ari Atolls is 150 fathoms, and that between the north and south Nillandoo Atolls is 200 fathoms in depth), that it is impossible to look at a map of them without believing that they were once more intimately related. And in this same archipelago, Mahlos-Mahdoo Atoll is divided by a bifurcating channel from 100 to 132 fathoms in depth, in such a manner, that it is scarcely possible to say whether it ought strictly to be called three separate atolls, or one great atoll not yet finally divided.

I will not enter on many more details; but I must remark that the curious structure of the northern Maldiva Atolls receives (taking into consideration the free entrance of the sea through their broken margins) a simple explanation in the upward and outward growth of the corals, originally based both on small detached reefs in their lagoons, such as occur in common atolls, and on broken portions of the linear marginal reef, such as bounds every atoll of the ordinary form. I cannot refrain from once again remarking on the singularity of these complex structures—a great sandy and generally concave disk rises abruptly from the unfathomable ocean, with its central expanse studded, and its edge symmetrically bordered with oval basins of coral-rock just lipping the surface of the sea, sometimes clothed with vegetation, and each containing a lake of clear water!

One more point in detail: as in two neighbouring archipelagoes corals flourish in one and not in the other, and as so many conditions before enumerated must affect their existence, it would be an inexplicable fact if, during the changes to which earth, air, and water are subjected, the reef-building corals were to keep alive for perpetuity on any one spot or area. And as by our theory the areas including atolls and barrier-reefs are subsiding, we ought occasionally to find reefs both dead and submerged. In all reefs, owing to the sediment being washed out of the lagoon or lagoon-channel to leeward, that side is least favourable to the long-continued vigorous growth of the corals; hence dead portions of reef not unfrequently occur on the leeward side; and these, though still retaining their proper wall-like form, are now in several instances sunk several fathoms beneath the surface. The Chagos group appears from some cause, possibly from the subsidence having been too rapid, at present to be much less favourably circumstanced for the growth of reefs than formerly: one atoll has a portion of its marginal reef, nine miles in length, dead and submerged; a second has only a few quite small living points which rise to the surface; a third and fourth are entirely dead and submerged; a fifth is a mere wreck, with its structure almost obliterated. It is remarkable that in all these cases, the dead reefs and portions of reef lie at nearly the same depth, namely, from six to eight fathoms beneath the surface, as if they had been carried down by one uniform movement. One of these "half-drowned atolls," so called by Capt. Moresby (to whom I am indebted for much invaluable information), is of vast size, namely, ninety nautical miles across in one direction, and seventy miles in another line; and is in many respects eminently curious. As by our theory it follows that new atolls will generally be formed in each new area of subsidence, two weighty objections might have been raised, namely, that atolls must be increasing indefinitely in number; and secondly, that in old areas of subsidence each separate atoll must be increasing indefinitely in thickness, if proofs of their occasional destruction could not have been adduced. Thus have we traced the history of these great rings of coral-rock, from their first origin through their normal changes, and through the occasional accidents of their existence, to their death and final obliteration.

In my volume on *"Coral Formations"* I have published a map, in which I have coloured all the atolls dark-blue, the barrier-reefs pale-blue, and the fringing-reefs red. These latter reefs have been formed whilst the land has been stationary, or, as appears from the frequent presence of upraised organic remains, whilst it has been slowly rising: atolls and barrier-reefs, on the other hand, have grown up during the directly opposite movement of subsidence, which movement must have been very gradual, and in the case of atolls so vast in amount as to have buried every mountain-summit over wide ocean-spaces. Now in this map we see that the reefs tinted pale and dark blue, which have been produced by the same order of movement, as a general rule manifestly stand near each other. Again we see that the areas with the two blue tints are of wide extent; and that they lie separate from extensive lines of coast coloured red, both of which circumstances might naturally have been inferred, on the theory of the nature of the reefs having been governed by the nature of the earth's movement. It deserves notice, that in more than one instance where single red and blue circles approach each other, I can show that there have been oscillations of level; for in such cases the red or fringed circles consist of atolls, originally by our theory formed during subsidence, but subsequently upheaved; and on the other hand, some of the pale-blue or encircled islands are composed of coral-rock, which must have been uplifted to its present height before that subsidence took place, during which the existing barrier-reefs grew upwards.

Authors have noticed with surprise, that although atolls are the commonest coral-structures throughout some enormous oceanic tracts, they are entirely absent in other seas, as in the West Indies: we can now at once perceive the cause, for where there has not been subsidence, atolls cannot have been formed; and in the case of the West Indies and parts of the East Indies, these tracts are known to have been rising within the recent period. The larger areas, coloured red and blue, are all elongated; and between the two colours there is a degree of rude alternation, as if the rising of one had balanced the sinking of the other. Taking into consideration the proofs of recent elevation both on the fringed coasts and on some others (for instance, in South America) where there are no reefs, we are led to conclude that the great continents are for the most part rising areas; and from the nature of the coral-reefs, that the central parts of the great oceans are sinking areas. The East Indian Archipelago, the most broken land in the world, is in most parts an area of elevation, but surrounded and penetrated, probably in more lines than one, by narrow areas of subsidence.

I have marked with vermilion spots all the many known active volcanoes within the limits of this same map. Their entire absence from every one of the great subsiding areas, coloured either pale or dark blue, is most striking; and not less so is the coincidence of the chief volcanic chains with the parts coloured red, which we are led to conclude have either long remained stationary, or more generally have been recently upraised. Although a few of the vermilion spots occur within no great distance of single circles tinted blue, yet not one single active volcano is situated within several hundred miles of an archipelago, or even small group of atolls. It is, therefore, a striking fact that in the Friendly Archipelago, which consists of a group of atolls upheaved and since partially worn down, two volcanoes, and perhaps more, are historically known to have been in action. On the other hand, although most of the islands in the Pacific which are encircled by barrier-reefs, are of volcanic origin, often with the remnants of craters still distinguishable, not one of them is known to have ever been in eruption. Hence in these cases it would appear, that volcanoes burst forth into action and become extinguished on the same spots, according as elevatory or subsiding movements prevail there. Numberless facts could be adduced to prove that upraised organic remains are common wherever there are active volcanoes; but until it could be shown that in areas of subsidence, volcanoes were either absent or inactive, the inference, however probable in itself, that their distribution depended on the rising or falling of the earth's surface, would have been hazardous. But now, I think, we may freely admit this important deduction.

Taking a final view of the map, and bearing in mind the statement made with respect to the upraised organic remains, we must feel astonished at the vastness of the areas, which have suffered changes in level either downwards or upwards, within a period not geologically remote. It would appear, also, that the elevatory and subsiding movements follow nearly the same laws. Throughout the spaces interspersed with atolls, where not a single peak of high land has been left above the level of the sea, the sinking must have been immense in amount. The sinking, moreover, whether continuous, or recurrent with intervals sufficiently long for the corals again to bring up their living edifices to the surface, must necessarily have been extremely slow. This conclusion is probably the most important one which can be deduced from the study of coral formations;—and it is one which it is difficult to imagine, how otherwise could it ever have been arrived at. Nor can I quite pass over the probability of the former existence of large archipelagoes of lofty islands, where now only rings of coral-rock scarcely break the open expanse of the sea, throwing some light on the distribution of the inhabitants of the other high islands, now left standing so immensely remote from each other in the midst of the great oceans. The reef-constructing corals have indeed reared and preserved wonderful memorials of the subterranean oscillations of level; we see in each barrier-reef a proof that the land has there subsided, and in each atoll a monument over an island now lost. We may thus, like unto a geologist who had lived his ten thousand years and kept a record of the passing changes, gain some insight into the great system by which the surface of this globe has been broken up, and land and water interchanged.

QUESTIONS AND SUGGESTIONS

1. Darwin viewed coral reefs with the eye of reason. He probably would not have come to conclusions like these had he not been familiar with the work of Charles Lyell; see Stephen Toulmin and June Goodfield's discussion of Lyell's work in *The Discovery of Time*. Write an essay describing something in nature and analyzing the historical influences which produced its empirical features.

2. Compare Darwin on coral reefs with Michael Faraday on candles. What accounts for the differences between the two essays, which both have a class of objects as their subject?

3. Darwin typically concludes his exposition of complicated material with a restrained but deeply sincere expression of awe at what he has observed. Thus, his last statement about seeing with the perspective of a geologist who has lived for ten thousand years. How does this compare with Rachel Carson's or Lewis Thomas's style of expressing emotion?

4. Darwin enumerates three classes of coral reefs and describes members of each. But he presents several conflicting theories of reef formation before presenting his own observations; from this, he develops the theory of subsidence as one which explains all three classes. His theory has been proved correct, except that we now know that rather than the whole sea floor sinking, the islands press down upon the sea floor with their weight, creating a saucer-like depression. But would Darwin's theory have convinced you if you had been a member of the audience to whom he read it in 1837?

MARIAN C. DIAMOND

The Impact of Air Ions

MARIAN C. DIAMOND (b. 1926) is a professor of anatomy at the University of California at Berkeley, who has been researching the neurology of rats since 1977. The excerpt that follows is from *Enriching Heredity* (1988). She has held visiting lectureships in Australia and China and has served on the boards of the National Science Foundation, the National Institutes of Health, and the World Health Organization. Her primary research interests are in the development and function of the brain.

The possibility that physical environmental stimuli, such as electromagnetic waves, may have effects on the brain has already received some attention by other investigators. That the balance of positive and negative ions in the atmosphere may be another such stimulus is suggested by the early work of Sulman et al. These investigators studied the responses of groups of "weather-sensitive" individuals to changes in the ionization of air. Though the relationship between air ions and weather-sensitive people is not well established, scientists are attempting to gain a better understanding of human sensitivity to weather and of the biological effects of atmospheric ions in general. According to Sulman, only about 30% of the population is weather-sensitive.

Being one of these weather-sensitive people, I found this developing field to be of great interest. As a girl, walking two miles home from school uphill through the sagebrush, I could always predict an oncoming thunderstorm by the presence of painful headaches. How could the weather cause my head to hurt? How could something in the atmosphere alter something within the skull? I like the passage in *Notes from the Underground* in which Dostoevsky carries out a discussion with his readers on whether "science itself will teach man that he never had any will of his own, . . . that everything he does is not done by his willing it, but is done by it-

239

self, by the laws of nature." The laws of nature do sometimes act in unexpected ways, as we shall learn in this chapter by studying the responses of the cerebral cortex to the levels of air ions.

The word "ion" was coined by Michael Faraday because ions migrate. Ionization occurs in the air from collisions between particles. Air ions form when energy from radioactive compounds in the soil or from cosmic rays acts upon a gas molecule and causes it to eject an electron. The molecule stripped of its electron becomes a positive ion, and the displaced electron then attaches to a neighboring molecule, which becomes a negative ion. Each molecular ion immediately attracts nonionized molecules to form a cluster of 8 to 12 molecules—an air ion. Negative ions can be produced naturally around falling fresh water, e.g., waterfalls or showers. In normal, clean air over land there are about 1500 to 4000 ions per cubic centimeter, but in polluted air in cities or in stale air in buildings, the ion content can be considerably lower. The ion content in the building where I work, the Life Sciences Building, was measured to have fewer than 100 positive ions per cubic centimeter.

In 1960 Krueger and Smith suggested that the known physiological and bicohemical effects of air ions may be due to their ability to alter the metabolism of biogenic amines, of which the neurotransmitter serotonin is one. There is evidence that the cyclic nucleotides participate in some of the metabolic events underlying synaptic transmission within the brain and that changes in the brain content of compounds such as adenosine $3',5'$-monophosphate (cyclic AMP) and guanosine $3',5'$-monophosphate (cyclic GMP) may reflect interactions between some neurotransmitters and their synaptic receptors. We wondered whether air ions could influence serotonin, cyclic AMP, as well as cyclic GMP in our rat brains and, if so, were these changes related to behavior?

The neurotransmitter serotonin has been implicated in changes in mood. A well-controlled study correlating air ions, serotonin concentration in the urine, and mood was performed by Sigel at the University of California in San Francisco. She asked each of 33 men to spend two hours in a small room containing high levels of either positive or negative ions; later, each man spent two hours in a room with the other kind of ion. Sigel found that both types of ions reduced serotonin and made the men feel good. The results indicated that ions clearly influence behavior, but in a complex way that can only be understood one step at a time.

We decided to take one such step in our laboratory. Since no single investigator was adequately prepared to produce and monitor the ions, to design the behavioral paradigm, and to quantify the chemical changes, we formed a team of scientists: Elaine Orenberg, Ph.D., a neurochemist at Stanford University; Michael Yost, a graduate student in public health at Berkeley; Professor Albert Krueger, our specialist in the field of air ions; James R. Connor, a graduate student in the physiology and anatomy department at Berkeley; Michael Bissell, M.D., a neuropathologist at the Veterans Administration hospital in Martinez, California; and myself.

Our experiments were divided into three groups. The first study, which used pups as subjects, was undertaken to investigate whether the effects of negative air ions depended on whether the animals lived in enriched or impoverished environments; serotonin and cerebral cortical weight were measured.

In the first experiment, groups of male Long-Evans rats were housed in enriched or impoverished environments, with and without increased levels of negative air ions. Nine littermate pairs of 6-day-old pups were distributed among six mothers into two experimental environments: a multifamily enriched condition and a unifamily impoverished condition. (These behavioral conditions were similar to those reported in connection with our study of the effects of differential environments on the developing brain in animals not yet weaned.) In the multifamily enriched condition, three mothers, each with three pups, were housed in one large cage filled with toys. In the unifamily condition, one mother was housed with her three pups in a standard colony sized cage with no toys. There were three cages housing the unifamily rats.

To serve as controls animals lived in atmospheric conditions, i.e., received air delivered by the building ventilation system and contained fewer than 100 positive ions per cubic centimeter of air. These animals were grouped in a similar fashion and were housed in either a wire-mesh enrichment cage or standard laboratory cages (18 pups in the atmospheric-condition groups), but in a separate room. All pups lived in their respective housing from the age of 6 days to the age of 26 days.

Both the enriched and the nonenriched animals exposed to nega-

tive air ions lived in Lucite cages, with Lucite toys in the enrichment cage. A grounded wire-mesh floor was suspended over the
sawdust waste collection tray at the cage bottom. A fan supplied
continually moving air to each cage, and a filter was used to free
the air of particulates.

An air ion density of 1×10^5 negative ions per cubic centimeter
was maintained in both the large and small cages. Negative air
ions were generated by corona discharge from Amcor Modulion
power supplies and regulated in each cage by adjusting the ionization potential with separate variable transformers on the ac lines
of each generator. The ion density of the air in each cage was
measured with the aid of a Royco volumetric counter and was correlated with the flow of current from the wire-mesh floor to an
earth ground. Daily checks of the current flow to ground were
made to ensure proper operation of the ionization equipment.

Before brain samples were dissected for chemical and wet weight
measures, all animals were coded to prevent experimental bias.
Uniform samples of somatosensory and occipital cortices were sur-
gically removed from both hemispheres, weighed, and frozen. All
procedures were accomplished within four minutes.

The results of these experiments were most revealing, informing
us of the many ways the ion content of the air can affect the
cerebral cortex. First, in both the initial and the replication experiments the wet weights obtained before the chemical assays
from the samples of the somatosensory and occipital cortices were
heavier in rats receiving negative ions than in those living in atmospheric conditions. Here was evidence that exposure of rats to
high levels of negative ions increased the weight of the outer
layers of the brain.

The chemical changes were equally consistent. We learned that
the enriched rats living in a negative ion atmosphere had significantly less serotonin (61%, $p < 0.01$) and cyclic AMP (45%,
$p < 0.05$) in the somatosensory cortex than the enriched rats living
in atmospheric conditions. Cyclic AMP is a second messenger; i.e.,
it translates extracellular messages into an intracellular response.
Serotonin and cyclic AMP concentrations in the occipital cortex
were also significantly less (45%, $p < 0.05$, and 35%, $p < 0.05$, respectively) in the enriched rats receiving negative ions than in the
enriched rats living in atmospheric conditions. It appears that in

the enriched condition, but not in the impoverished condition, negative ions prevent the increases in serotonin and cyclic AMP concentrations that occur in atmospheric conditions. The cyclic GMP levels increased slightly, though not significantly, in the somatosensory cortex and in the occipital cortex of both the enriched and impoverished rats receiving negative ions in comparison with their counterparts living in atmospheric conditions.

Our data on serotonin are consistent with other reports showing that negative air ions decrease brain serotonin. Gilbert used negative ions to reduce emotionality caused by isolation; the reduction in emotionality paralleled a decrease in serotonin. Olivereau found that brief exposure of rats to 1.5×10^5 negative ions per cubic centimeter of air modified their ability to adapt to a stressful situation. He considered the effect to be due to a reduction in serotonin caused by the action of air ions.

In most instances, the pattern of change in the content of cyclic AMP was in the same direction as the change in serotonin. If changes in cyclic AMP are serotonin-dependent, cyclic AMP might reflect the metabolic alteration in serotonin resulting from negative air ions, as well as from the living situation.

The content of cyclic GMP was relatively unchanged by either the atmospheric or the environmental state. This may indicate that the neuronal cells selectively control steady-state tissue levels of the two cyclic nucleotides by independent regulatory mechanisms.

We chose six-day-old rats as experimental animals in this study in order to identify potential effects of air ions on neural development. There appears to be a serotonin-dependent adenylate cyclase system which participates in some of the metabolic events underlying synaptic transmission in very young animals, and which decreases in sensitivity with age. Our results showing that serotonin and cyclic AMP are similar in their response patterns to our environmental conditions may be due to this coupling. It is conceivable, however, that negative air ions could have shifted the apparent developmental stage of the rats in this study by changing the concentration of some hormones, such as prolactin, sex hormones, or serotonin-derived melatonin. A direct effect of air ions on prolactin levels was proposed by Olivereau in his study of the effects of air ions on the spontaneous movements of amphibian larvae. A direct hormonal effect of this kind in our system may have caused the increased cortical weights and decreased cortical levels of serotonin and the second messenger cyclic AMP reported here. A further study comparing the effects of negative and positive ions and different environments on various neurotransmitters would help clarify how the responses change the developmental stage of the animal brain. Undoubtedly, the direction of change of one transmitter does not indicate how other transmitters are altered.

Having shown these effects of negative ions on weight and neurotransmitter concentration in the young, developing cortex, we turn to our second study, one measuring the effects of elevated levels of negative ions in older brains. For this experiment we were interested in learning whether brains from animals living for an extended period of time in conditions with high levels of negative ions aged more rapidly than those from animals in atmospheric conditions. It seemed of importance to understand the effects of long-term exposure to these ions because of evidence which suggests that negative ions increase metabolic activity. For example, it has been shown that negative ions increase ciliary motility in the respiratory tract. In addition, as was found in the first experiment described in this chapter, the weight of the developing cerebral cortex increases with high levels of negative ions. With these two pieces of information, it appeared to us that negative ions might have the effect of accelerating maturation and aging by somehow increasing metabolic processes. If this were the case, could the ions affect the aging process of nerve cells? In order to answer this question, we planned an experiment with Professor George Ellman in the psychiatry department at the University of California in San Francisco to learn whether the aging pigment called lipofuscin, found in nerve and glial cells and thought to be a metabolic byproduct, could be measurably altered with prolonged use of negative ions.

For this experiment, female Long-Evans rats, seven months of age, were separated into one of four environmental conditions: (1) an enriched condition with the addition of 1×10^5 negative air ions per cubic centimeter (eight rats in a large cage, with the toys changed every other day excluding weekends); (2) a nonenriched condition with 1×10^5 negative air ions per cubic centimeter (two rats per small cage, providing a total of six rats in three small

cages); (3) an enriched condition with atmospheric air with the
rats caged as in the first group; and (4) a nonenriched condition
with atmospheric air, with the rats caged as in the second group.
At 14 months of age, seven months later, the somatosensory and
occipital cortices of the left hemisphere were removed, weighed,
and frozen in order to assay for the lipofuscin.

Since the results were statistically nonsignificant, they are mentioned only to show trends. Quite unexpectedly we learned that
the lipofuscin concentration was less in the enriched animals than
in the nonenriched, whether the animals were exposed to excess
ions or not. The percentage differences in lipofuscin concentration
between the enriched and nonenriched were quite large. With the
negative ions, the differences were 16%, and in atmospheric conditions, 9%. The standard errors of the mean were much larger
than those seen in the wet weights of the same tissues, for example. These data indicate that there is great variability in lipofuscin
accumulation in individual rats during aging. They also suggest
that animals living in an enriched condition do not accumulate
lipofuscin at the same rate as animals living in a nonenriched condition. As has been mentioned previously, enriched rats have more
large capillaries in their cortices than nonenriched rats. It is possible that the more efficient vascular system transports precursors
of lipofuscin away from the cell. If so, then we still do not know
the answer to our original question, Does lipofuscin accumulate
faster in enriched or in nonenriched brains? Whichever way the
process does work, the data suggest there is less of the pigment in
the more active enriched brains when the animals have lived in
their respective conditions for as long as seven months during
early adulthood. Since it is thought that an accumulation of lipofuscin in the cytoplasm of the cell can hinder its function, less pigment in the enriched cells could be interpreted as beneficial.

Though our primary interest in air ions was their effect on the
brain, it was relatively easy to take blood samples and examine
the effects of high negative ion concentrations on white blood cell
counts, as well as on brain sections taken from the same animals.
If we could find changes in the white blood cells, some of which
are related to antibody–antigen responses, then we would be encouraged to utilize the brain slices to make the more tedious counts
of glial cells, which also have antigen markers. In an experiment

using mice in three separate age groups, we found that white blood cell counts were altered with high levels of negative ions. The design of this third experiment was as follows. One-month-old mice lived in standard colony mouse cages with 25 mice per cage. One cage, placed on an electrically grounded floor, was supplied with negative ions (2×10^3 ions per cubic centimeter), and the other cage had no electric field. The animals were exposed to these conditions for different time periods: 3 weeks, 2 months, and $3\frac{1}{2}$ months. For the two groups of mice and for the three different age groups, the lymphocyte counts were greater by 11% ($p < 0.01$) to 18% ($p < 0.05$) in the ion-rich condition than in the animals living in the grounded cages. The neutrophil count was in the opposite direction: 25% ($p < 0.01$) to 36% ($p < 0.05$) lower in the ion-rich atmosphere than in the other group. These results clearly indicate that the number of white blood cells can be changed with high levels of negative ions. With these results we will some day return to our slides of brain sections to examine the ion effect on glial cell populations.

These experiments have informed us how sensitive nerve and blood cells are to different levels of atmospheric ions. In the brain cells, the strength of the effects depends on whether the animals are living in enriched or impoverished conditions. The high levels of negative ions decreased both serotonin and cyclic AMP in the enriched but not in the impoverished animals' brains. The ion studies have been used primarily to indicate how subtle changes in the ion content of the air can alter brain structure and chemistry, not to endorse or discriminate against the use of ions. Undoubtedly, further research will demonstrate that many other structures are equally sensitive to the ion content in the air.

QUESTIONS AND SUGGESTIONS

1. How would you characterize Diamond's prose style? How does she attempt to engage the audience in her research?
2. Detail the subjects, processes, and results of her studies.
3. Write an extended definition of the term "ion" or the term "serotonin" using specialized dictionaries or encyclopedias.

4. Research the effects of other physical stimuli such as cold and heat or pleasure and pain on the body and the brain. For an added dimension, you may wish to consult Stanley Milgram's *Obedience to Authority* (Harper and Row, 1974) to develop a discussion of responses.

5. Prepare a summary, abstract, and executive summary of this essay.

6. Diamond is the author of numerous articles on neurology. Compile an annotated bibliography of her works and analyze her style as a writer in a short essay.

JULIAN HUXLEY

Evolutionary Progress

SIR JULIAN HUXLEY (1887–1975) was from a great English family: his grand-father, T. H. Huxley, a famous defender of Darwin; his father, Leonard Huxley, a biographer and historian; his brother Aldous, a novelist; and a half-brother, Andrew, a Nobel laureate. Julian Huxley was a key figure in the in-troduction of experimental concepts and methods into biological research in England. For example, he applied the theory of axial gradients to obtain marked developmental modifications in frogs' eggs through temperature varia-tions. His work on genetic control of the rates of developmental processes opened up the whole field of developmental genetics. He stimulated the field of ecology with his studies of territoriality and ritualization in birds. *Evolution: The Modern Synthesis* (1942), of which this essay is the penultimate chapter, evolved from an address to professional zoologists in which Huxley called for a synthetic theory of evolution that would bring together research in genetics, developmental physiology, paleontology, and other related fields. Here, Huxley examines the implications of evolution for our concept of progress.

Is Evolutionary Progress a Scientific Concept?

The question of evolutionary or biological progress remains. There still exists a very great deal of confusion among biologists on the subject. Indeed the confusion appears to be greater among professional biologists than among laymen. This is probably due to the common human failing of not seeing the wood for the trees; there are so many more trees for the professional!

The chief objections that have been made to employing *progress* at all as a biological term, and to the use of its correlates *higher* and *lower* as applied to groups of organisms, are as follows. First, it is objected that a bacillus, a jellyfish, or a tapeworm is as well adapted to its environment as a bird, an ant, or a man; and that therefore it is incorrect to speak of the latter as higher than the former, and illogical to speak of the processes leading to their pro-duction as involving progress. An even simpler objection is to use mere survival as criterion of biological value, instead of adapta-tion. Man survives: but so does the tubercle bacillus. So why call man the higher organism of the two?

248

A somewhat similar argument points to the fact that evolution, both in the fossil record and indirectly, shows us numerous examples of specialization leading to increased efficiency of adaptation to this or that mode of life; but that many of such specialized lines become extinct, while most of the remainder reach an equilibrium and show no further change.

This type of objection, then, points to certain fundamental attributes of living things or their evolution, uses them as definitions of progress, and then denies that progress exists because they are found in all kinds of organisms, and not only in those that the believers in the existence of progress would call progressive.

A slightly less uncompromising attitude is taken up by those who admit that there has been an increase of complexity or an increase in degree of organization, but deny that this has any value, biological or otherwise, and accordingly refuse to dignify this trend by a term such as progress, with all its implications.

Some sociologists, faced with the problem of reconciling the objective criteria of the physical sciences with the value criteria with which the sociological data confronts them, take refuge in the ostrich-like attitude of refusing to recognize any scale of values. Thus Doob in a recent book (1940) writes:

> In this way, the anthropologist has attempted to remove the idea of progress from his discipline. For him, there is just change, or perhaps a tendency towards increasing complexity. Neither change nor complexity is good or bad; there are differences in degree, not in quality or virtue. . . . The sweep of historical progress reveals no progressive trend. . . .

By introducing certain objective criteria into our definition of progress, as we do in the succeeding section, this objection can be overcome, at least for pre-human evolution. In regard to human evolution, however, as we shall see in the concluding section of this chapter, the nettle must be grasped, and human values given a place among the criteria of human progress.

The second main type of objection consists in showing that many processes of evolution are not progressive in any possible sense of the word, and then drawing the conclusion that progress does not exist. For instance, many forms of life, of which the brachiopod *Lingula* is the best-known example, have demonstrably remained unchanged for enormous periods of several hundreds of millions of years; if a Law of Progress exists, the objectors argue, how is it that such organisms are exempt from its operations?

A variant of this objection is to draw attention to the numerous cases where evolution has led to degeneration involving a degradation of form and function, as in tapeworms, *Sacculina* and other parasites, in sea-squirts and other sedentary forms: how, it is asked, can the evolutionary process be regarded as progressive if it produces degeneration?

This category of objections can be readily disposed of. Objectors of this type have been guilty of setting up an Aunt Sally of their own creation for the pleasure of knocking her down. They have assumed that progress must be universal and compulsory: when they find, quite correctly, that universal and compulsory progress does not exist, they state that they have proved that progress does not exist. This, however, is an elementary fallacy. The task before the biologist is not to define progress *a priori*, but to proceed inductively to see whether he can or cannot find evidence of a process which can legitimately be called progressive. It may just as well prove to be partial as universal. Indeed, human experience would encourage search along those lines; the fact that man's progress in mechanical arts, for instance, in one part of the world is accompanied by complete stagnation or even retrogression in other parts, is a familiar fact. Thus evolution may perfectly well include progress without being progressive as a whole.

The first category of objections, when considered closely, is seen to rest upon a similar fallacy. Here again an Aunt Sally has been set up. Progress is first defined in terms of certain properties: and then the distribution of those properties among organisms is shown not to be progressive.

These procedures would be laughable, if they were not lamentable in arguing a lack of training in logical thought and scientific procedure among biologists. Once more, the elementary fact must be stressed that the only correct method of approach to the problem is an inductive one. Even the hardened opponents of the idea of biological progress find it difficult to avoid speaking of higher and lower organisms, though they may salve their consciences by putting the words between inverted commas. The unprejudiced observer will accordingly begin by examining various types of "so-called higher" organisms and trying to discover what characters they possess in common by which they differ from "lower" organisms. He will then proceed to examine the course of evolution as recorded in fossils and deduced from indirect evidence, to see what the main types of evolutionary change have been; whether some of them have consistently led to the development of characters diagnostic of "higher" forms; which types of change have been most successful in producing new groups, dominant forms, and so forth. If evolutionary progress exists, he will by this means discover its factual basis, and this will enable him to give an objective definition.

The Definition of Evolutionary Progress

Proceeding on these lines, we can immediately rule out certain characters of organisms and their evolution from any definition of biological progress. Adaptation and survival, for instance, are universal, and are found just as much in "lower" as in "higher" forms: indeed, many higher types have become extinct while lower ones have survived. Complexity of organization or of life-cycle cannot be ruled out so simply. High types *are* on the whole more complex than low. But many obviously low organisms exhibit remarkable complexities, and, what is more cogent, many very complex types have become extinct or have speedily come to an evolutionary dead end.

Perhaps the most salient fact in the evolutionary history of life is the succession of what the paleontologist calls dominant types. These are characterized not only by a high degree of complexity for the epoch in which they lived, but by a capacity for branching out into a multiplicity of forms. This radiation seems always to be accompanied by the partial or even total extinction of competing main types, and doubtless the one fact is in large part directly correlated with the other.

In the early Paleozoic the primitive relatives of the Crustacea known as the trilobites were the dominant group. These were succeeded by the marine arachnoids called sea-scorpions or eurypterids, and they in turn by the armoured but jawless vertebrates, the ostracoderms, more closely related to lampreys than to true fish. The fish, however, were not far behind, and soon became the dominant group. Meanwhile, groups both from among the arthropods and the vertebrates became adapted to land life, and towards the close of the Paleozoic, insects and amphibians could both claim the title of dominant groups. The amphibia shortly gave rise to the reptiles, much more fully adapted to land life, and the primitive early insects produced higher types, such as beetles, hymenoptera and lepidoptera. Higher insects and reptiles were the dominant land groups in the Mesozoic, while among aquatic forms the fish remained preeminent, and evolved into more efficient types: from the end of the Mesozoic onwards, however, they show little further change.

Birds and mammals began their career in the Mesozoic, but only became dominant in the Cenozoic. The mammals continued their evolution through the whole of this epoch, while the insects reached a standstill soon after its beginning. Finally man's ancestral stock diverged, probably towards the middle of the Cenozoic, but did not become dominant until the latter part of the Ice Age.

In these last two cases, the rise of the new type and the downfall of the old was without question accompanied and facilitated by world-wide climatic change, and this was probably true for other biological revolutions, such as the rise of the reptiles to dominance.

When the facts concerning dominant groups are surveyed in more detail, they yield various interesting conclusions. In the first place, biologists are in substantial agreement as to what were and what were not dominant groups. Secondly, some groups once dominant have become wholly extinguished, like the trilobites, eurypterids and ostracoderms, while others survive only in a much reduced form, many of their sub-groups having been extinguished, as with the reptiles or the monotremes, or their numbers enormously diminished, as with the larger non-human placentals. Those which do not show reduction of one or the other sort have remained to all intents and purposes unchanged for a longer or shorter period of geological time, as with the insects or the birds. Finally, later dominant groups normally arise from an unspecialized line of an earlier dominant group, as the birds and reptiles from among the early reptiles, man from the primates among the mammals. . . . They represent, in fact, one among many lines of adaptive radiation; but they differ from the others in containing the potentiality of evolving so as to become dominant on a new level, with the aid of new properties. Usually the new dominance is marked by a fresh outburst of radiation: the only exception to this rule is Man, a dominant type which shows negligible radiation of the usual structurally-adapted sort, but makes up for its absence by the complexity of his social life and his division of labour.

If we then try to analyse the matter still further by examining the characters which distinguish dominant from non-dominant and earlier from later dominant groups, we shall find first of all, efficiency in such matters as speed and the application of force to overcome physical limitations. The eurypterids must have been better swimmers than the trilobites, the fish, with their muscular tails, much better than either; and the later fish are clearly more efficient aquatic mechanisms than the earlier. Similarly the earlier reptiles were heavy and clumsy, and quite incapable of swift running. Sense-organs also are improved, and brains enlarged. In the latest stages the power of manipulation is evolved. Through a combination of these various factors man is able to deal with his environment in a greater variety of ways, and to apply greater forces to its control, than any other organism.

Another set of characteristics concerns the internal environment. Lower marine organisms have blood or body-fluids identical in saline concentrations with that of the seawater in which they live; and if the composition of their fluid environment is changed, that of their blood changes correspondingly. The higher fish, on the other hand, have the capacity of keeping their internal environment chemically almost constant. Birds and mammals have gone a step further: they can keep the temperature of their internal environment constant too, and so are independent of a wide range of external temperature change.

The early land animals were faced with the problem of becoming independent of changes in the moisture-content of the air. This was accomplished only very partially by amphibia, but fully by adult reptiles and insects through the development of a hard impermeable covering. The freeing of the young vertebrate from dependence on water was more difficult. The great majority of amphibians are still aquatic for the earlier part of their existence: the elaborate arrangements for rendering the reptilian egg cleidoic

. . . were needed to permit of the whole life-cycle becoming truly terrestrial.

There is no need to multiply examples. The distinguishing characteristics of dominant groups all fall into one or other of two types—those making for greater control over the environment, and those making for greater independence of the environment. Thus advance in these respects may provisionally be taken as the criterion of biological progress.

The Nature and Mechanism of Evolutionary Progress

It is important to realize that progress, as thus defined, is not the same as specialization. Specialization, as we have previously noted, is an improvement in efficiency of adaptation for a particular mode of life: progress is an improvement in efficiency of living in general. The latter is an all-round, the former a one-sided advance. We must also remember that in evolutionary history we can and must judge by final results. And there is no certain case on record of a line showing a high degree of specialization giving rise to a new type. All new types which themselves are capable of adaptive radiation seem to have been produced by relatively unspecialized ancestral lines.

Looked at from a slightly different angle, we may say that progress must in part at least be defined on the basis of final results. These results have consisted in the historical fact of a succession of dominant groups. And the chief characteristic which analysis reveals as having contributed to the rise of any one of these groups is an improvement that is not one-sided but all-round and basic. Temperature-regulation, for instance, is a property which affects almost every function as well as enabling its possessors to extend their activities in time and their range in space. Placental reproduction is not only a greater protection for the young—a placental mother, however hard-pressed, cannot abandon her unborn embryo—but this additional protection, together with the later period of maternal care, makes possible the extension of the plastic period of learning which then served as the basis for the further continuance of progress.

It might, however, be held that biological inventions such as the lung and cleidoic shelled egg, which opened the world of land to the vertebrates, are after all nothing but specializations. Are they not of the same nature as the wing which unlocked the kingdom of the air to the birds, or even to the degenerations and peculiar physiological changes which made it possible for parasites to enter upon that hitherto inaccessible habitat provided by the intestines of other animals? This is in one sense true; but in another it is untrue. The bird and the tapeworm, although they did conquer a new section of the environment, in so doing were as a matter of actual fact cut off from further progress. Theirs was only a special-ization, though a large and notable one. The conquest of the land, however, not only did not involve any such limitations, but made demands upon the organism which could be and in some groups were met by further changes of a definitely progressive nature. Temperature-regulation, for instance, could never have arisen through natural selection except in an environment with rapidly-changing temperatures: in the less changeable waters of the sea the premium upon it would not be high enough.

Of course a progressive advance may eventually come to a dead end, as has happened with the insects, when all the biological pos-sibilities inherent in the type of organization have been exploited. From one point of view it might be permissible to call such a trend a long-range specialization; but it would appear more rea-sonable to style it a form of progress, albeit one which is destined eventually to be arrested. It is limited as opposed to unlimited progress.

A word is needed here on the restricted nature of biological progress. We have seen that evolution may involve downward or lateral trends, in the shape of degeneration or certain forms of specialization, and may also leave certain types stable. Further, lower types may persist alongside higher, even when the lower are representatives of a once-dominant group that includes the higher types. From this, it will first be seen, as we already mentioned, that progress is not compulsory and universal; and secondly that it will not be so marked in regard to the average of biological effi-ciency as to its upper limit. Progress, in other words, can most readily be studied by examining the *upper levels* of biological efficiency (as determined by our criteria of control and independ-ence) attained by life at successive periods of its evolution.

For this, during the earlier part of life's history, we must rely

The Past Course of Evolutionary Progress

One somewhat curious fact emerges from a survey of biological progress as culminating for the evolutionary moment in the dominance of *Homo sapiens*. It could apparently have pursued no other general course than that which it has historically followed: or, if it be impossible to uphold such a sweeping and universal negative, we may at least say that among the actual inhabitants of the earth, past and present, no other lines could have been taken which would have produced speech and conceptual thought, the features that form the basis for man's biological dominance.

Multicellular organization was necessary to achieve the basis for adequate size: without triploblastic development and a blood-system, elaborate organization and further size would have been impossible. Among the coelomates, only the vertebrates were eligible as agents for unlimited progress, for only they were able to achieve the combination of active efficiency, size, and terrestrial existence on which the later stages of progress were of necessity based. Only in the water have the molluscs achieved any great advance. The arthropods are not only hampered by their necessity for moulting; but their land representatives, as was first pointed out by Krogh, are restricted by their tracheal respiration to very small size. They are therefore also restricted to cold-bloodedness and to a reliance on instictive behaviour. . . . Lungs were one needful precursor of intelligence. Warm blood was another, since only with a constant internal environment could the brain achieve stability and regularity for its finer functions. This limits us to birds and mammals as bearers of the torch of progress. But birds were ruled out by their depriving themselves of potential hands in favour of actual wings, and perhaps also by the restriction of their size made necessary in the interests of flight.

Remain the mammals. During the Tertiary epoch, most mammalian lines cut themselves off from the possibility of ultimate progress by concentrating on immediate specialization. A horse or a lion is armoured against progress by the very efficiency of its limbs and teeth and sense of smell: it is a limited piece of organic machinery. As Elliot Smith has so fully set forth, the penultimate steps in the development of our human intelligence could never have been taken except in arboreal ancestors, in whom the fore-limb could be converted into a hand, and sight inevitably became the dominant sense in place of smell. But, for the ultimate step, it was necessary for the anthropoid to descend from the trees before he could become man. This meant the final liberation of the hand, and also placed the evolving creature in a more varied environment, in which a higher premium was placed upon intelligence. Further, the foetalization necessary for a prolonged period of learning could only have occurred in a monotocous species. . . . Weidenreich (1941) maintains that the attainment of the erect posture was a necessary prerequisite for the final stages in human cerebral evolution.

The last step yet taken in evolutionary progress, and the only one to hold out the promise of unlimited (or indeed of any further) progress in the evolutionary future, is the degree of intelligence which involves true speech and conceptual thought: and it is found exclusively in man. This, however, could only arise in a monotocous mammal of terrestrial habit, but arboreal for most of its mammalian ancestry. All other known groups of animals, except the ancestral line of this kind of mammal, are ruled out. Conceptual thought is not merely found exclusively in man: it could not have been evolved on earth except in man.

Evolution is thus seen as a series of blind alleys. Some are extremely short—those leading to new genera and species that either remain stable or become extinct. Others are longer—the lines of adaptive radiation within a group such as a class or sub-class, which run for tens of millions of years before coming up against their terminal blank wall. Others are still longer—the lines that have in the past led to the development of the major phyla and their highest representatives; their course is to be reckoned not in tens but in hundreds of millions of years. But all in the long run have terminated blindly. That of the echinoderms, for instance, reached its climax before the end of the Mesozoic. For the arthropods, represented by their highest group, the insects, the full stop seems to have come in the early Cenozoic: even the ants and bees have made no advance since the Oligocene. For the birds, the Miocene marked the end; for the mammals, the Pliocene.

Only along one single line is progress and its future possibility being continued—the line of man. If man were wiped out, it is in the highest degree improbable that the step to conceptual thought would again be taken, even by his nearest kin. In the ten or twenty million years since his ancestral stock branched off from the rest of the anthropoids, these relatives of his have been forced into their own lines of specialization, and have quite left behind them that more generalized stage from which a conscious thinking creature could develop. Although the irreversibility of evolution is not an impossibility *per se,* it is probably an actual impossibility in a world of competing types. Man might conceivably cause the capacity for speech and thought to develop by long and intensive selection in the progeny of chimpanzees or gorillas; but Nature, it seems certain, could never do so.

One of the concomitants of organic progress has been the progressive cutting down of the possible modes of further progress, until now, after a thousand or fifteen hundred million years of evolution, progress hangs on but a single thread. The thread is the human germ-plasm. As Villiers de l'Isle-Adam wrote in *L'Ève Future,* "L'Homme . . . seul, dans l'univers n'est pas fini."

Progress in the Evolutionary Future

What of the future? In the past, every major step in evolutionary progress has been followed by an outburst of change. For one thing the familiar possibilities of adaptive radiation may be exploited anew by a number of fresh types which dominate or extinguish the older dispensation by the aid of the new piece of organic machinery which they possess. Or, when the progressive step has opened up new environmental realms, as was the case with lungs and the shelled egg, these are conquered and peopled; or the fundamental progressive mechanism may itself be improved, as was the case with temperature-regulation or the pre-natal care of the young in mammals.

Conscious and conceptual thought is the latest step in life's progress. It is, in the perspective of evolution, a very recent one, having been taken perhaps only one or two and certainly less than ten million years ago. Although already it has been the cause of many and radical changes, its main effects are indubitably still to come. What will they be? Prophetic phantasy is a dangerous pastime for a scientist, and I do not propose to indulge it here. But at least we can exclude certain possibilities. Man, we can be certain, is not within any near future destined to break up into separate radiating lines. For the first time in evolution, a new major step in biological progress will produce but a single species. The genetic variety achieved elsewhere by radiating divergence will with us depend primarily upon crossing and recombination. . . .

We can also set limits to the extension of his range. For the planet which he inhabits is limited, and adventures to other planets or other stars are possibilities for the remote future only.

During historic times, all or almost all of the increase in man's control over nature have been non-genetic, owing to his exploitation of his biologically unique capacity for tradition, whereby he is provided with a modificational substitute for genetic change. The realization of the possibilities thus available will continue to play a major part in human evolution for a very long period, and may contribute largely to human progress.

More basic, however, though much slower in operation, are changes in the genetic constitution of the species, and it is evident that the main part of any large genetic change in the biologically near future must then be sought in the improvement of the fundamental basis of human dominance—the feeling, thinking brain, and the most important aspect of such advance will be increased intelligence, which, as A. Huxley . . . has stressed, implies greater disinterestedness and fuller control of emotional impulse.

First, let us remind ourselves that . . . we with our human type of society must give up any hope of developing such altruistic instincts as those of the social insects. It would be more correct to say that this is impossible so long as our species continues in its present reproductive habits. If we were to adopt the system advocated by Muller (1936) and Brewer (1937), of separating the two functions of sex—love and reproduction—and using the gametes of a few highly endowed males to sire all the next generation, or if we could discover how to implement the suggestion of Haldane in his *Daedalus* and reproduce our species solely from selected germinal tissue-cultures, then all kinds of new possibilities would emerge. True castes might be developed, and some at least of them might be endowed with altruistic and communal impulses. In any case, as A. Huxley (1937) points out in an interesting discussion, progress (or, I would prefer to say, future human progress) is dependent on an increase of intraspecific co-operation until it preponderates over intraspecific competition.

Meanwhile there are many obvious ways in which the brain's level of performance could be genetically raised—in acuteness of perception, memory, synthetic grasp and intuition, analytic capacity, mental energy, creative power, balance, and judgment. If for all these attributes of mind the average of our population could be raised to the level now attained by the best endowed ten-thousandth or even thousandth, that alone would be of far-reaching evolutionary significance. Nor is there any reason to suppose that such quantitative increase could not be pushed beyond its present upper limits.

Further, there are other faculties, the bare existence of which is as yet scarcely established: and these too might be developed until they were as commonly distributed as, say, musical or mathematical gifts are to-day. I refer to telepathy and other extrasensory activities of mind, which the painstaking work of Rhine (1935), Tyrrell (1935), and others is now forcing upon the scientific world as a subject demanding close analysis.

If this were so, it would be in a sense only a continuation of a process that has already been at work—the utilization by man for his own ends of hitherto useless by-products of his mental constitution. The earlier members of the Hominidae can have had little use for the higher ranges of aesthetic creation or appreciation, for mathematics or pure intellectual construction. Yet to-day these play a large part in human existence, and have come to possess important practical consequences as well as value in and for themselves. The development of telepathic knowledge or feeling, if it really exists, would have equally important consequences, practical as well as intrinsic.

In any case, one important point should be borne in mind. After most of the major progressive steps taken by life in the past, the progressive stock has found itself handicapped by characteristics developed in earlier phases, and has been forced to modify or abandon these to realize the full possibilities of the new phase (see M. Roberts, 1920, 1930, for various examples of forced adjustment to new conditions, but with the caveat that some are highly speculative, and that all are presented in a lamarckian frame of reference which often obscures their true significance). This evolutionary fact is perhaps most obvious in relation to the vertebrates' emergence from water on to land. But it applies in other cases too. The homothermy of mammals demanded the scrapping of scales and the substitution of hair; man's erect posture brought with it a number of anatomical inconveniences. But man's step to conscious thought is perhaps more radical in this respect than any other.

By means of this new gift, man has discovered how to grow food instead of hunting it, and to substitute extraneous sources of power for that derived from his own muscles. And for the satisfaction of a few instincts, he has been able to substitute new and more complex satisfactions, in the realm of morality, pure intellect, aesthetics, and creative activity.

The problem immediately poses itself whether man's muscular power and urges to hunting prowess may not often be a handicap to his new modes of control over his environment, and whether some of his inherited impulses and his simpler irrational satisfactions may not stand in the way of higher values and fuller enjoyment. The poet spoke of letting ape and tiger die. To this pair, the cynic later added the donkey, as more pervasive and in the long run more dangerous. The evolutionary biologist is tempted to ask whether the aim should not be to let the mammal die within us, so as the more effectually to permit the man to live.

Here the problem of values must be faced. Man differs from any previous dominant type in that he can consciously formulate values. And the realization of these in relation to the priority determined by whatever scale of values is adopted, must accordingly be added to the criteria of biological progress, once advance has reached the human level. Furthermore, the introduction of such criteria based upon values, in addition to the simpler and more objective criteria of increasing control and independence which sufficed for pre-human evolution, alters the direction of progress. It might perhaps be preferable to say that it alters the level on which progress occurs. True human progress consists in increases of aesthetic, intellectual, and spiritual experience and satisfaction.

Of course, increase of control and of independence is necessary for the increase of these spiritual satisfactions; but the more or less measurable and objective control over and independence of external environment are now merely subsidiary mechanisms serving as the material basis for the human type of progress, and the really significant control and independence apply to man's mental states—his control of ideas to give intellectual satisfaction, of form and colour or of sound to give aesthetic satisfaction, his independence of inessential stimuli and ideas to give the satisfaction of mystic detachment and inner ecstasy.

The ordinary man, or at least the ordinary poet, philosopher and theologian, is always asking himself what is the purpose of human life, and is anxious to discover some extraneous purpose to which he and humanity may conform. Some find such a purpose exhibited directly in revealed religion; others think that they can uncover it from the facts of nature. One of the commonest methods of this form of natural religion is to point to evolution as manifesting such a purpose. The history of life, it is asserted, manifests guidance on the part of some external power; and the usual deduction is that we can safely trust that same power for further guidance in the future.

I believe this reasoning to be wholly false. The purpose manifested in evolution, whether in adaptation, specialization, of biological progress, is only an apparent purpose. It is just as much a product of blind forces as is the falling of a stone to earth or the ebb and flow of the tides. It is we who have read purpose into evolution, as earlier men projected will and emotion into inorganic phenomena like storm or earthquake. If we wish to work towards a purpose for the future of man, we must formulate that purpose ourselves. Purposes in life are made, not found.

But if we cannot discover a purpose in evolution, we can discern a direction—the line of evolutionary progress. And this past direction can serve as a guide in formulating our purpose for the future. Increase of control, increase of independence, increase of internal co-ordination; increase of knowledge, of means for co-ordinating knowledge, of elaborateness and intensity of feeling—those are trends of the most general order. If we do not continue them in the future, we cannot hope that we are in the main line of evolutionary progress any more than could a sea-urchin or a tapeworm.

As further advice to be gleaned from evolution there is the fact we have just discussed, that each major step in progress necessitates scrapping some of the achievements of previous advances. But this warning remains as general as the positive guidance. The precise formulation of human purpose cannot be decided on the basis of the past. Each step in evolutionary progress has brought new problems, which have had to be solved on their own merits; and with the new predominance of mind that has come with man, life finds its new problems even more unfamiliar than usual. This last step marks a critical point in evolution, and has brought life into situations that differ in quality from those to which it was earlier accustomed.

The future of progressive evolution is the future of man. The future of man, if it is to be progress and not merely a standstill or a degeneration, must be guided by a deliberate purpose. And this human purpose can only be formulated in terms of the new attributes achieved by life in becoming human. Man, as we have stressed, is in many respects unique among animals: his purpose must take account of his unique features as well as of those he shares with other life.

Human purpose and the progress based upon it must accordingly be formulated in terms of human values; but it must also take account of human needs and limitations, whether these be of a biological order, such as our dietary requirements or our mode of reproduction, or of a human order, such as our intellectual limitations or our inevitable subjection to emotional conflict.

Obviously the formulation of an agreed purpose for man as a whole will not be easy. There have been many attempts already. To-day we are experiencing the struggle between two opposed ideals—that of the subordination of the individual to the community, and that of his intrinsic superiority. Another struggle still in progress is between the idea of a purpose directed to a future life in a supernatural world, and one directed to progress in the existing world. Until such major conflicts are resolved, humanity can have no single major purpose, and progress can be but fitful and slow. Before progress can begin to be rapid, man must cease being afraid of his uniqueness, and must not continue to put off the responsibilities that are really his on to the shoulders of mythical gods or metaphysical absolutes. . . .

But let us not forget that it is possible for progress to be achieved. After the disillusionment of the early twentieth century it has become as fashionable to deny the existence of progress and to brand the idea of it as a human illusion, as it was fashionable in the optimism of the nineteenth century to proclaim not only its existence but its inevitability. The truth is between the two extremes. Progress is a major fact of past evolution; but it is limited to a few selected stocks. It may continue in the future but it is not inevitable; man, by now become the trustee of evolution, must work and plan if he is to achieve further progress for himself and so for life.

This limited and contingent progress is very different from the *deus ex machina* of nineteenth-century thought, and our optimism may well be tempered by reflection on the difficulties to be overcome. None the less, the demonstration of the existence of a general trend which can legitimately be called progress, and the definition of its limitations, will remain as a fundamental contribution of evolutionary biology to human thought.

QUESTIONS AND SUGGESTIONS

1. Huxley writes in a very well-organized way. He covers an enormous amount of material by limiting each section so that it bears a carefuly determined relationship to the whole essay. Reveal his organizational pattern by making an outline of the essay.

2. Do you find Huxley's concept of evolutionary progress convincing? What other concepts of progress exist, and how valid are they? Why are values so important in a concept of progress?

3. Clearly, the concept of evolutionary progress was debated at the time Huxley wrote, as it continues to be. Why does Huxley make note of others' positions at the same time that he argues his own? Write an essay taking one side in a current scientific debate.

GREGORY BATESON

The Messages of Nature and Nurture

GREGORY BATESON (1904–1980) is best known for labeling and defining the "double bind," a linguistic action in which a command is issued then the speaker's tone of voice indicates it is wrong to follow the command. Yet, his hypothesis that the "double bind" contributes to schizophrenia in children has been superseded by more recent research emphasizing the genetic and biochemical causes of the disease. Born in Cambridge, England, in 1904, Bateson was the son of the biologist William Bateson and Beatrice Durham. In 1936, Bateson married the anthropologist Margaret Mead. Their daughter Mary Catherine Bateson is an author, anthropologist, and linguist. During his sixty-year career, Bateson was a professor at the University of Sydney, Australia, Columbia University, and Harvard University. From 1964 through 1972, he was Chief of Biological Relations at the Oceanic Institute of Hawaii, and in 1977 he began his tenure as a Regent of the University of California. He was a fellow of the American Association for the Advancement of Science and, in 1946, a Guggenheim Fellow. His research interests included philosophy, evolution, schizophrenia, and learning and culture. Among his many books were *Naven: A Study of the Problems Suggested by a Composite Picture of the Culture of a New Guinea Tribe Drawn from Three Points of View* (1936); *The Balinese Character* (1943); *Communication: The Social Matrix of Psychiatry* (with Jurgen Ruesch) (1950/1987); *Steps to an Ecology of a Mind* (1975); and *Angels Fear: Towards an Epistemology of the Sacred* (1987). Bateson, who died in San Francisco in 1980, was the subject of two books by his daughter, *About Bateson: Essays on Gregory Bateson* (1977) and *With a Daughter's Eye: A Memoir of Margaret Mead and Gregory Bateson* (1984).

I am continually surprised at the lighthearted way in which scientists assert that some characteristic of an organism is to be explained by invoking *either* the environment *or* the genotype. Let me make clear, therefore, what I believe to be the relation between these two explanatory systems. It is precisely in the *relation* between the two systems that the tangles occur which make me hesitate to assign any given characteristic to one or the other.

In the description of an organism, consider any component that is subject to change under environmental impact—for example, the color of the skin. In those human beings who are not albinos the color of the skin is subject to darkening or tanning when the

WRITING FOR PROFESSIONAL AUDIENCES

skin is exposed to sunshine. Now, imagine that we are asking about a particular human being, to what is the particular degree of brown in his skin to be assigned? Is it genotypically or phenotypically determined?

The answer, of course, will involve both genotype and phenotype. Some persons are born browner than others, and all, so far as I know (with the exception of albinos), are capable of becoming more brown under sunshine. We may, therefore, say immediately that genotype is involved in two ways: in determining the starting point of tanning and in determining an ability to tan. On the other hand, the environment is involved in exploiting the ability to tan, to produce the phenotypic color of the given individual.

The next question is whether there is not only a tanning under the influence of sunshine but also, possibly, an *increase in ability* to tan under sunshine. Could we, by tanning and bleaching and tanning and bleaching an individual successively a number of times, increase his "skill" in turning brown under sunshine? If so, then the genotype and the phenotype are both involved at the next level of abstraction, the genotype in providing the individual with not only an ability to tan but with an ability to learn to tan and the environment correspondingly taking up this ability.

But, then again, there is the question of whether the genotype conceivably provides an ability *to learn to change the ability* to tan. This would seem exceedingly unlikely, but the question must be asked when we are dealing with a creature subject to learning, to environmental impact. Insofar as the creature is subject to such impact, it is always made so by the characteristics of its genotypic determination.

In the end, if we want to ask about tanning or about any other phenomenon of environmental change or learning, the question we have to ask is the logical type of the specification provided by the genotype. Does it define skin color? Does it define ability to change skin color? Does it define the ability to change the ability to change skin color? And so on. For every descriptive proposition we may utter about a phenotype, there is a background of explanation which, at successive logical type levels, will always peel off into the genotype. The particular environment, of course, is still always relevant for explanation.

I believe that something of this sort is necessarily so, and from this it follows that a major question or set of questions we have to be aware of concerns the logical typing of the genotypic message.

The case of blood sugar is interesting. The actual concentration of sugar in the blood varies from minute to minute with intake of carbohydrates, liver action, exercise, and the length of time between meals, etc. But these changes must be kept within tolerance. There is an upper threshold and a lower threshold and the organism *must* keep the blood sugar level within these limits, on pain of extreme discomfort and/or death. But these limits are changeable under environmental pressure, such as chronic starvation, training, and acclimation. Finally, the abstract component in the trait—that indeed blood sugar has an upper limit that is modifiable by experience—must be referred to genetic control.

It is said that in the 1920s when Germany was restricted by the Treaty of Versailles to a parade-ground army of ten thousand men, very strict tests were applied to the men who volunterd for this army. They were to be the cream of the rising generation not only in physique but also in physiology and dedication. A blood sample was taken from each volunteer at the beginning of the test. He was then asked to climb over a simple barrier in the recruiting office, and then to climb back, and then to go on climbing until he could not climb anymore. When *he* decided that he "could not," his blood was again sampled. Those were accepted into the army who were most able to reduce their blood sugar, overcoming exhaustion by determination. No doubt the trait *"able to reduce blood sugar"* would be subject to quantitative change through training or practice, but also, no doubt, some individuals would (probably for genetic reasons) respond more and more rapidly to such training.

It is no simple matter to identify the trait that is specified by the genotype. Let us consider some cases at a rather naive level. There used to be, in the American Museum of Natural History in New York, an exhibit designed to show the bell-shaped curve of random distribution of a variable. This curve was made from a bucket of clams randomly collected on the Long Island shore. The clams in question have a variable number of ridges going from the hinge radially towards the periphery of the shell. The number of ridges varied, as I remember, from about nine to ten to about twenty. A curve was made by piling up one shell on top of the other—all the nine-ridged shells to make one vertical column and then next to it all the ten-ridged shells—and then drawing a curve on the wall behind them at the height of the different piles. It appeared then that somewhere in the middle range one column was higher than all the other columns and that the height of the columns fell off both towards the shells with fewer ridges and towards those with more ridges. But curiously, interestingly, the curve so produced was actually not a clean Gaussian curve. It was skewed. And it was in fact skewed so that the norm was closer to the end having fewer ridges.

I looked at this curve and wondered why it was skewed, and it occurred to me that perhaps the coordinates were wrongly chosen. That perhaps what affected the growth of the clam was not the number of ridges but how closely packed the ridges might be. That is, there might be more difference from the growing clam's point of view between having nine ridges and having ten ridges than there is between having eighteen ridges and having nineteen. How much space is there for more ridges? What angle does each ridge occupy? It therefore followed that perhaps the curve should have been plotted not against the number of ridges but against the reciprocal of this number. Against, that is, the average angle between ridges. If these increments had been used, the curve would have undergone a change, because in fact the curve of the reciprocal $(x = 1/y)$ is not a straight line. It is a parabola. Therefore, a skewing of the curve might be neutralized. It was clear in any case that if the curve plotted against the number of ridges were in fact a normal curve, then the curve plotted against the angles between ridges could not possibly be one. Conversely, if the curve plotted against the angle should be normal, then the other, plotted against the number of ridges, would have to be skewed. It therefore seemed a reasonable question to ask whether one curve might perhaps give in some sense a *truer* picture of the state of affairs than the other— "truer" in the sense that one curve rather than the other might more accurately reflect the genotypic message.

A very little arithmetic and some graph paper showed that the curve was immediately much less skewed and, in a sense, would have been a better museum exhibit had it been plotted against the angle rather than the number of ridges.

Thinking about this very simple example will illustrate what I mean by inquiring about the logical type of the genotypic message. Is it conceivable that in the case of these clams the genotypic message contains somehow a direct reference to a *number* of ridges? Or is it more probable that in fact the message contains no substantive of that order at all—probably contains no analog of a substantive? There might indeed be no "word" for an angle, so that the whole message would be somehow carried as a name of an operation, if you please, in some sort of group-theoretic definition of the pattern of ridges and angles. In this case, estimating by angles (that is, by the *relation* between ridges) will certainly be a more appropriate way to describe the organism than stating the number of ridges.

We, after all, can look at the whole clam and count the ridges, but in the process of growth the message of the DNA must be locally read. A reference to number cannot be *locally* useful, but a reference to *relation* between the local patch of tissue and the neighboring regions could conceivably be significant. The larger patterns must always be carried in the form of detailed instructions to the component parts.

The essence of the matter is that if we are concerned with environment and with genotypic determination, what we ultimately most desire is that our description of the individual phenotype shall be in a language appropriate to the genotypic messages and to the environmental impacts that have shaped that phenotype.

If we look at, say, a crab, we note that it has two chelae and eight walking appendages on the thorax, i.e. two claws and four pairs of legs. But it is not a trivial matter to decide whether we will say this animal has "ten appendages on the thorax" or "five pairs of appendages on the thorax" or "one pair of claws and four pairs of walking legs on the thorax." No doubt there are other ways of stating the matter, but the point I want to make is merely that one of these ways can be better than others in applying to the phenotype a syntax of description that will reflect the messages from the genotype that have determined that phenotype. And note that the description of phenotype that best reflects the injunctions of genotype will also, necessarily, bring to the fore whatever components of the phenotype have been determined by environmental impact. The two sorts of determinism will, in fact, be sorted out and their relations clearly indicated in the ultimate perfect description, which will reflect both.

But note further that the notion of number represented in our description of the clams is a totally different concept—a different logical type, if you please—from number as represented by the number of appendages on the thorax of the crab. In the one case, that of the clam, number would seem to refer to a quantity. The very fact that it varies on something like a normal curve indicates immediately that this is a matter of more-or-less. On the other hand, the appendages of the crab are strictly limited and are not a quantity but essentially a pattern. And this difference between quantity and pattern is important right through the biological world and no doubt right through the behavior of the entities in the biological world. We may expect it to be not only anatomical but also behavioral. (I presume that the theoretical approaches for anatomy, physiology, and behavior are a single set of approaches.)

In the relation between genetics and morphogenesis, we face over and over again problems that are really double problems. This double character of almost every problem in communication was summarized by Warren McCulloch in the title of his famous paper "What is a number, that a man may know it, and a man, that he may know a number?"* In our case the problems become: "What is the message of the DNA that the embryo may receive it, and what is an embryo, that it may receive the DNA's message?" The problem becomes sharply compelling when we deal with such matters as symmetry, metamerism (segmentation), and multiple organs.

I suggested above that we might say either that the crab has "ten" appendages or "five pairs" of appendages on its thorax. But this evidently won't do at all. Imagine that a particular locality—a small region—of the developing embryo must receive some version of the instructions governing these multiple limbs. The version that specifies "five" or "ten" can be of no use to the specific restricted locality. How shall this spot—this bunch of cells in this particular region of the developing embryo—know about a number that is in fact embodied in the larger aggregate of tissues developing those appendages? The restricted locality of the single cell

---

* Repr. in his *Embodiments of Mind* (Cambridge, Mass.: MIT P, 1965), 1–18.

needs to have information not only that there should be five appendages but that there are already elsewhere four or three or two or whatever the situation at a given moment may be. The situation is left obscure even when we change from saying "five" or "ten" to saying a *pattern* of five, a quincunx, say, or whatever may be appropriate.

What the particular patch of growing cells or the single cell needs is *both* information about the total pattern *and* information about what the particular patch is to become in this gestalt. (Alternatively the whole embryology must be what is called mosaic. In the smaller nematodes, it is claimed that the outcome of every cell division is literally pre-scribed.)

The relation between pattern and quantity becomes especially important when we look at the relations between environmntal impact and genotypic determinants. Another way of looking at the difference between these two sorts of explanation is that genotypic explanation commonly invokes the digital and the patterned, whereas the impacts of environment are likely to take the form of quantities, stresses, and the like. We may think, perhaps, of the soma—the developing body—of the individual as the arena where quantity meets pattern. And precisely because environmental determination tends to be quantitative, while genotypic determination tends to be a matter of patterns, people—scientists—exhibit strong preferences and are ready to guess lightheartedly at which explanation shall be applied in which case. It would seem that some people prefer quantitative explanations, while others prefer explanations by the invoking of pattern.

These two states of mind, which are almost different enough to be regarded as different epistemologies, have become embodied in political doctrines. Specifically, Marxist dialectic has concerned itself with the relation between quantity and quality or, as I would say, between quantity and pattern. The orthodox notion is, as I understand it, that all important social change is brought on or precipitated by quantitative "pressures," "tensions," and the like. These quantities supposedly build up to some sort of breaking point, at which point a discontinuous step occurs in the social evolution leading to a new gestalt. The essence of the matter is that it is quantity that determines this step and, by corollary, it is assumed that the necessary ingredients for the next change will always and necessarily be present when the quantity becomes sufficiently stressful.

A model commonly cited from the physical world for this phenomenon is that of a chain which under tension will always, in the end, break at its weakest link. In case the links are equal and virtually nondiscriminable, the chain may go beyond its ordinary breaking point but, in the end, a weakest link will be discovered and this will be the point of fracture. Another model is the crystallization of a slowly cooling liquid. The crystallization process will always start from some particular point—some minute inequality or fragment of dirt—and once started will proceed to completion. Very pure and clean substances in very smooth containers may be supercooled by a few degrees, but in the end the change will occur and, if there was supercooling, it is likely to be rapid.

Another example of relationship between quantitative and qualitative change is the relation between jamming of traffic on the roads and the population of automobiles on which such jamming depends. The population of automobiles in a given region grows slowly through the years, but the velocity at which automobiles can travel remains constant until a certain threshold value of automobile population is reached. The curve of the population of automobiles is a slowly increasing, smooth curve accelerating slightly as time passes. In contrast, the curve of time spent by each automobile on each mile of the road runs along at a horizontal constant value to a certain point. Then quite suddenly, when the number of automobiles passes that threshold, there is jamming on the roads and the curve representing time spent per mile on the road skyrockets steeply.

We may say that an increase in number of automobiles was in some sense "good"—had a positive value up to a certain point—but beyond that point the number of automobiles in the area has become toxic.

The dialectical view of history is comparable. It assumes that at a given point in history, say the middle of the nineteenth century, social differentiation and pressures were such that a theory of evolution of a particular kind would be generated to reflect that social system. To the Marxist it is, as I understand it, irrelevant whether that theory was produced by Darwin, or Wallace, or Chambers, or by any other of the half-dozen leading biologists who were then on the point of creating an evolutionary theory of that general kind. The Marxist assumption is that when the time is ripe the man will always be present. There will always be someone who will form the crystallization point for the new gestalt. And, indeed, the theory of evolution and its history would seem to confirm this. There were several men in the 1850s who were ready to create a theory of evolution, and this theory, or something like it, was more or less inevitable, give or take ten years before or after the actual date of publication of *The Origin of Species*. It was also, no doubt, politically inevitable that Lamarckian evolution should disappear from the scene at that time and that the cybernetic philosophy of evolution, though actually proposed by Wallace, should not become a dominant theme.

For the Marxists the essence of the matter is that quantity will determine what happens and that pattern will be generated in response to quantitative change.

My own view of the matter, as it has developed in the last few years, is almost the precise contrary of this or, should we say, the precise complement of this. Namely, I have argued that quantity can never under any circumstances explain pattern. That the informational content of quantity, as such, is zero.

It has seemed increasingly clear to me that the vulgar use of "energy" as an explanatory device is fallacious precisely because quantity does *not* determine pattern. I would argue that quantity of tension applied to the chain will not break it *except* by the discovery of the weakest link—that, in fact, the pattern is latent in the chain before the application of the tension and is, as the photographers might say, "developed" under tension.

I am thus temperamentally and intellectually one of those who prefer explanation by pattern to explanation by quantity.

Recently, however, I have begun to see how these two sorts of explanation may fit together. I have for a long time felt an uneasiness about what is meant by the concept of "a question," about whether it was possible for something like a question to be embodied in the prelinguistic biological world.

Let me be clear that I don't now mean a question that a perceiving organism might put to an environment. We might say that the rat exploring in a box is in some sense asking if that box is safe or dangerous, but that is not what concerns us here.

Instead, I am asking whether, at a deeper level, there can be something like a *question* expressed in the language of the injunctions, etc., that are at the base of genetics, morphogenesis, adaptation, and the like. What would the word "question" mean at this deep biological level?

The paradigm I have been carrying in my mind for some time to represent what I mean by a "question" at the morphogenetic level is the sequence of events that follows fertilization of the vertebrate egg, as demonstrated with the eggs of the frog. The unfertilized frog egg, as is well known, is a radially symmetrical system in which the two poles (the upper or "animal" and the lower or "vegetal") are differentiated in that the animal pole has more protoplasm and indeed is the region of the nucleus, while the vegetal pole is more heavily endowed with yolk. But the egg is, it seems, similar all around its equator. There is no differentiation of the plane that will be the future plane of bilateral symmetry of the tadpole. This plane is then determined by the entry of a spermatozoon, usually somewhat below the equator, so that a line drawn through the point of entry and connecting the two poles defines the future midventral line of bilateral symmetry. The environment thus provides the *answer* to the question: Where? which seems to be latent all around the unfertilized egg.

In other words, the egg does not contain the needed information, and neither is this information embodied in any complex way in the DNA of the spermatozoon. Indeed, with a frog's egg, a spermatozoon is not even necessary. The effect can be achieved by pricking the egg with the fiber of a camel's-hair brush. Such an unfertilized egg will then develop into a fully grown frog, albeit haploid (having only a half number of chromosomes).

It was this figure that I carried in mind as a paradigm for thinking about the nature of a question. It seemed to me that we might think of the state of the egg immediately before fertilization as a state of question, a state of *readiness to receive a certain piece of information,* information that is then provided by the entry of the spermatozoon.

Combining this model with what I have said about quantity and quality in the Marxist dialectic and relating all this to the battles that have been fought to and fro over the problem of environmental determinism versus genetic determinism and to the Lamarckian battles, which have also had their political angles, it occurred to me that perhaps *the question is quantitative while the answer is qualitative.* It seemed to me that the state of the egg at the moment of fertilization could probably be described in terms of some quantity of "tension," tension that is in some sense resolved by the essentially digital or qualitative answer provided by the spermatozoon. The question: Where? is a distributed quantity. The answer "there" is a precise digital answer—a digital patterned resolution.

To go back to the chain and the weakest link, it seemed to me that both in the chain and in the case of the frog's egg, the particular digital answer is provided out of the random. The question, however, comes in from the quantitative, represented by increasing tension.

Let us return to the problem of describing an organism and of what is going to happen to the parts of that description as the creature undergoes the processes of growth, environmental impact, or evolution. We may conveniently follow Ashby in regarding the description of an organism as a list of propositional variables, running perhaps to some millions of propositions. Each of these propositions or values has the characteristic that above a certain level it will become lethal. In other words, the organism has the task of maintaining every variable within limits of tolerance—upper and lower. The organism is enabled to do this by the fact of homeostatic circuits. The variables of which we have a list are very densely and complexly interconnected in circuits having homeostatic or metahomeostatic characteristics. In such systems there are two types of pathology, or should we say pathways to disaster. In the first place, any monotone change—i.e. any continuous increase or decrease in the value of any variable—must inevitably lead to destruction of the system or to such deep (or "radical") disturbance that it is almost impossible to say that we are now dealing with the "same" system. That is one pathway to disaster, death, or radical change.

On the other hand, it is equally disastrous to peg or fix the value of any variable, because fixing the value of any variable will in the end disrupt the homeostatic processes. If the variable is one that normally changes rather easily and quickly, the fixing of it will tend to disturb those slow-moving variables which are at the core of the whole organism. The acrobat, for example, is unable to maintain his position on the high wire if the position of his balancing pole relative to himself is fixed. He must vary the one in order to maintain the truth of the ongoing propositional variable: "I am on the wire."

The picture we get is that a qualitative change in any variable is going to have a discontinuous effect upon the homeostatic structure. What was said earlier about quantity and quality becomes an alternative version of what Ashby has said in describing the system as a series of homeostatic circuits. These are partly synonymous descriptions. Ashby has added a new facet in pointing out that to *prevent* change in the superficial variables is to *promote* change in the more profound. (This is the process that is exploited in the strategy of the "obedience strike," when protesting workers achieve a slowdown simply by conforming strictly to regulations.)

The matter becomes much more complex when we are talking not about evolution, with changes that occur once and for all, but about embryology. In the process of development, a lot of crises are going to occur having this Ashbian form. Perhaps merely growth in size would be enough to disrupt a whole series of homeostases. The embryology is then going to have step functions built into it because of this curious relation between quantity and pattern. Furthermore, the embryo usually cannot trust to the random to provide the specification of the fracture planes of the system where it will break under the stress of some continuous change. In evolution, the random must be trusted to provide the planes of fracture, but in embryology these patterns of breaking must themselves be reliably determined by DNA message or by some other circumstance within the carefully protected embryo. As a crab automatically breaks off its own legs, or a lizard its tail, so there must be a place, a fracture plane, ready to define the break in embryological process and the new gestalt following the break.

The whole matter becomes still more complicated when we start to think of response or learning mediated by the central nervous system. A neuron, after all, is an almost precise analog of the frog's egg discussed earlier. A neuron is a component in the fabric of the organism that builds up a defined state of readiness—a quantity of "tension"—to be triggered by some external event, or by some external condition that can be made into an event. That state of tension is a "question" in the same sense that the tension of the frog's egg, previous to the arrival of the spermatozoon, is a question. The neuron, however, must go through the cycle over and over again and is, in fact, a specially designed piece of the organism that can do this over and over again. The neuron builds up a state, is triggered, and then builds up the state again. Both neurons and muscle fibers have this general characteristic. Upon this the whole organization of the creature depends.

This "relaxation-oscillation" characteristic of the neuron can be seen, if you will, as a repetitive, patterned sequence of "revolutionary" changes. (I understand that something of the sort occurs in parts of South America.)

Up to this point, I have focused upon *discontinuity* in the relation between input and response and have suggested that there are deep necessities lying behind the empirical facts of threshold and discontinuity. It is not only that the signal is improved by a high signal/noise ratio, it is also a necessity of complex cybernetic organization that many changes shall have flip-flop characteristics. Homeostatic control, however quantitative and "analogic" it may be, must always depend upon thresholds, and there will always be discontinuity between quantitative control and the breakdown of control that occurs when quantities become too great.

Let me now indicate another necessity—namely, that where discontinuity is lacking or is blurred by statistical response of smaller units (e.g. populations of neurons), regularities like those described by the Weber-Fechner laws shall operate. If there be cybernetic systems on other planets so complex that we might be willing to call them "organisms," then, surely, those systems must be characterized by a Weber-Fechner relationship whenever the relation across an interface is continuously variable on both sides.

What is asserted by the Weber-Fechner laws can be said in two ways:

1. Wherever a sense organ is used to compare two values of the same perceivable quantity (weight, brightness, etc.), there will be a threshold of perceivable difference below which the sense organ cannot discriminate between the quantities. This threshold of difference will be a *ratio,* and this ratio will be constant over a wide range of values. For example, if the experimental subject can just

discriminate between the perceived weight of thirty grams and the perceived weight of forty grams (a ratio of 3 : 4), then he will also just discriminate between three pounds and four pounds.

2. This is another way of saying that there will be a relation between input and sensation such that the quantity or intensity of sensation will vary as the logarithm of the intensity of the input.

This relation seems to characterize the interfaces between environment and nerve wherever the interface is mediated by a sense organ. It is especially precise in the case of the retina, as shown long ago by Selig Hecht.

Interestingly, the same relation that characterizes afferent, or incoming, impulses was encountered by Norbert Wiener at the interface between efferent nerve and muscle:* The isometric tension of the muscle is proportional to the logarithm of the frequency of neural impulses in the nerve serving that muscle.

So far as I know, there is no quantitative knowledge yet available of the relation between the response of an individual cell and the intensity of hormonal or other chemical messages impacting upon it. We do not know whether hormonal communication is Weber-Fechnerian.

As to the necessity of this Weber-Fechner relation in biological communication, the following considerations can be urged:

1. All digital information is concerned with *difference*. In map-territory relations (of whatever kind, in the widest sense) that which gets from the territory to the map is always and necessarily *news of difference*. If the territory is homogeneous, there is no mark upon the map. A succinct definition of information is "a difference which makes a difference at a distance."

2. The concept of difference enters *twice* into understanding the process of perception: first, there must be a difference latent or implicit in the territory, and secondly, that difference must be converted into an *event* within the perceiving system—i.e. the difference must overcome a threshold, must be different from a threshold value.

* Personal communication. I think that Wiener never published the details of this finding. He refers to it, however, in the introduction to his *Cybernetics*.

3. The sense organs are like the lining of the stomach in functioning as filters to protect the organism from the violence or toxicity of the environment. They must both admit the "news" and keep out the excessive impact. This is done by varying the response of the organ according to the intensity of the input. The logarithmic scale achieves precisely this: that the effect of inputs shall not increase according to their magnitude, but only according to the logarithm of their magnitude. The difference in effect between one hundred and one thousand units of input shall only equal the difference in effect between one and ten units.

4. The information the organism requires (which will confer survival value) matches the logarithmic scale. The organism benefits by very great sensitivity to very small impacts and does not need such precision in evaluating the gross. To hear a mouse in the grass or a dog bark a mile away—and still not be deafened by one's own voice shouting—that is the problem.

5. It seems that in all perception (not only in the biological) and in all measurement, there is something like a Weber-Fechner regularity. Even in man-made mechanical devices, the arithmetic sensitivity of the device falls off with the magnitude of the variable to be measured. A laboratory balance is only accurate in measuring comparatively small quantities, and error is usually computed as a percentage, i.e. as a *ratio*.

(Interestingly, the *Encyclopaedia Britannica* carries a skewed curve of the variable product of two hundred repetitions of a chemical process.* Here the skewing is probably a result of the Weber-Fechner law operating in the measurement of the various ingredients. Human eyes (and possibly some balances) are affected equally by equal *ratios*. They are therefore more sensitive to difference on the subtractive side of any norm, and less sensitive to difference on the additive side.)

6. It seems that the interface between nerve and environment is characterized by a deep difference in *kind*, i.e. in logical typing, between what is on one side of the interface and what is on the other. What is quantitative on the input side becomes qualitative and discontinuous on the perception side. Neurons obey an "all-or-nothing" rule, and, to make them report continuous variation in

* In a footnote to the 1965 edition, under "Error, theory of . . ."

a *quantity*, it is necessary to employ a statistical device—either the statistics of a population of neurons or the frequency of response of the single neuron.

All of these considerations work together to set the *mind* in special relation to the *body*. My arms and legs obey one set of laws and equations in terms of their purely physical characteristics— weight, length, temperature, etc. But, chiefly owing to the transformations of quantity imposed by the Weber-Fechner relation, my arms and legs obey quite different laws in their controlled motions within the communication systems I call "mind." We are dealing here with the interface between Creatura and Pleroma.

Fechner was surely a very remarkable man, at least a hundred years ahead of his time. He seems to have seen even then that the mind-body problem could not be settled by denying the reality of the mind. It was not good enough to assert that all biological causation was simply a materialistic impact of billiard balls. Nor was it good enough to assert that mind was a separate transcendent agent, a supernatural which could be divorced from body.

Fechner avoided both these forms of nonsense by asserting the logarithmic relation between message as carried in the communication systems of the body and the material quantities characterizing the impacts of the external "corporeal" universe.

Arguably it was Fechner who took the first steps, but much remains to be done. Our task over the next twenty years is to build up an Epistemology and a body of fact which shall unite the fields of genetics, morphogenesis, and learning. These three subjects are already clearly one field in which the concepts of a more abstract natural history or Epistemology will be explanatory themes. The Epistemology we shall construct will be both tautology—an abstract system making sense within its own terms—*and* natural history. It shall be that tautology onto which the empirical facts can be mapped. The metazoan cell, of course, already embodies just such an Epistemology. The relations between quantity and quality, the necessities of self-correction and homeostasis, and so on—all of this is determinant and component in the interaction between the cell and the environment it inhabits. But alas, the epistemologies of different human communities, notably those of the modern West, which govern interaction with the environment, are very far from providing what is needed.

QUESTIONS AND SUGGESTIONS

1. What function do the introductory examples serve? Describe the audience, objective, and thesis of this essay.

2. What role does Bateson assign to Marxist dialectic in his discussion of science?

3. Prepare extended definitions of the terms "genotype" and "phenotype."

4. What is the relationship between science and philosophy? Why do Western epistemologies fail to answer the questions of quantity and quality? You may wish to consult Aristotle and Plato as well as contemporary philosophers and other books by Bateson.

GARRETT HARDIN

The Tragedy of the Commons

GARRETT HARDIN (b. 1915) was trained as a biologist but later withdrew from his research on the culture of algae as a large-scale food source because he had come to believe that the production of more food aggravates rather than cures population problems. Since then, he has written widely on the ethical and social implications of science. *Biology: Its Human Implications* (1949, reedited in 1952) is the result of his work in updating college biology curricula. He is widely known as the author of this article, which first appeared in *Science* in 1968 and was based on a presidential address presented before the American Association for the Advancement of Science. Some of the ideas were advanced further in *Exploring New Ethics for Survival: The Voyage of the Spaceship Beagle,* which combines fiction and techniques of exposition.

At the end of a thoughtful article on the future of nuclear war, Wiesner and York[1] concluded that: "Both sides in the arms race are . . . confronted by the dilemma of steadily increasing military power and steadily decreasing national security. *It is our considered professional judgment that this dilemma has no technical solution.* If the great powers continue to look for solutions in the area of science and technology only, the result will be to worsen the situation."

I would like to focus your attention not on the subject of the article (national security in a nuclear world) but on the kind of conclusion they reached, namely that there is no technical solution to the problem. An implicit and almost universal assumption of discussions published in professional and semipopular scientific journals is that the problem under discussion has a technical solution. A technical solution may be defined as one that requires a change only in the techniques of the natural sciences, demanding little or nothing in the way of change in human values or ideas of morality.

In our day (though not in earlier times) technical solutions are always welcome. Because of previous failures in prophecy, it takes

[1] J. B. Wiesner and H. F. York, *Sci. Amer.* 211 (No. 4), 27 (1964).

courage to assert that a desired technical solution is not possible. Wiesner and York exhibited this courage; publishing in a science journal, they insisted that the solution to the problem was not to be found in the natural sciences. They cautiously qualified their statement with the phrase, "It is our considered professional judgment. . . ." Whether they were right or not is not the concern of the present article. Rather, the concern here is with the important concept of a class of human problems which can be called "no technical solution problems," and, more specifically, with the identification and discussion of one of these.

It is easy to show that the class is not a null class. Recall the game of tick-tack-toe. Consider the problem, "How can I win the game of tick-tack-toe?" It is well known that I cannot, if I assume (in keeping with the conventions of game theory) that my opponent understands the game perfectly. Put another way, there is no "technical solution" to the problem. I can win only by giving a radical meaning to the word "win." I can hit my opponent over the head; or I can drug him; or I can falsify the records. Every way in which I "win" involves, in some sense, an abandonment of the game, as we intuitively understand it. (I can also, of course, openly abandon the game—refuse to play it. This is what most adults do.)

The class of "No technical solution problems" has members. My thesis is that the "population problem," as conventionally conceived, is a member of this class. How it is conventionally conceived needs some comment. It is fair to say that most people who anguish over the population problem are trying to find a way to avoid the evils of over-population without relinquishing any of the privileges they now enjoy. They think that farming the seas or developing new strains of wheat will solve the problem—technologically. I try to show here that the solution they seek cannot be found. The population problem cannot be solved in a technical way, any more than can the problem of winning the game of tick-tack-toe.

Population, as Malthus said, naturally tends to grow "geometrically," or, as we would now say, exponentially. In a finite world this means that the per capita share of the world's goods must steadily decrease. Is ours a finite world?

A fair defense can be put forward for the view that the world is infinite; or that we do not know that it is not. But, in terms of the practical problems that we must face in the next generations with the foreseeable technology, it is clear that we will greatly increase human misery if we do not, during the immediate future, assume that the world available to the terrestrial human population is finite. "Space" is no escape.[2]

A finite world can support only a finite population; therefore, population growth must eventually equal zero. (The case of perpetual wide fluctuations above and below zero is a trivial variant that need not be discussed.) When this condition is met, what will be the situation of mankind? Specifically, can Bentham's goal of "the greatest good for the greatest number" be realized?

No—for two reasons, each sufficient by itself. The first is a theoretical one. It is not mathematically possible to maximize for two (or more) variables at the same time. This was clearly stated by von Neumann and Morgenstern,[3] but the principle is implicit in the theory of partial differential equations, dating back at least to D'Alembert (1717–1783).

The second reason springs directly from biological facts. To live, any organism must have a source of energy (for example, food). This energy is utilized for two purposes: mere maintenance and work. For man, maintenance of life requires about 1600 kilocalories a day ("maintenance calories"). Anything that he does over and above merely staying alive will be defined as work, and is supported by "work calories" which he takes in. Work calories are used not only for what we call work in common speech; they are also required for all forms of enjoyment, from swimming and automobile racing to playing music and writing poetry. If our goal is to maximize population it is obvious what we must do: We must make the work calories per person approach as close to zero as possible. No gourmet meals, no vacations, no sports, no music, no literature, no art. . . . I think that everyone will grant, without argument or proof, that maximizing population does not maximize goods. Bentham's goal is impossible.

In reaching this conclusion I have made the usual assumption that it is the acquisition of energy that is the problem. The ap-

[2] G. Hardin, *J. Hered.* 50, 68 (1959); S. von Hoernor, *Science* 137, 18 (1962).
[3] J. von Neumann and O. Morgenstern, *Theory of Games and Economic Behavior* (Princeton Univ. Press, Princeton, N.J., 1947), p. 11.

pearance of atomic energy has led some to question this assumption. However, given an infinite source of energy, population growth still produces an inescapable problem. The problem of the acquisition of energy is replaced by the problem of its dissipation, as J. H. Fremlin has so wittily shown.[4] The arithmetic signs in the analysis are, as it were, reversed; but Bentham's goal is still unobtainable.

The optimum population is, then, less than the maximum. The difficulty of defining the optimum is enormous; so far as I know, no one has seriously tackled this problem. Reaching an acceptable and stable solution will surely require more than one generation of hard analytical work—and much persuasion.

We want the maximum good per person; but what is good? To one person it is wilderness, to another it is ski lodges for thousands. To one it is estuaries to nourish ducks for hunters to shoot; to another it is factory land. Comparing one good with another is, we usually say, impossible because goods are incommensurable. Incommensurables cannot be compared.

Theoretically this may be true; but in real life incommensurables *are* commensurable. Only a criterion of judgment and a system of weighting are needed. In nature the criterion is survival. Is it better for a species to be small and hideable, or large and powerful? Natural selection commensurates the incommensurables. The compromise achieved depends on a natural weighting of the values of the variables.

Man must imitate this process. There is no doubt that in fact he already does, but unconsciously. It is when the hidden decisions are made explicit that the arguments begin. The problem for the years ahead is to work out an acceptable theory of weighting. Synergistic effects, nonlinear variation, and difficulties in discounting the future make the intellectual problem difficult, but not (in principle) insoluble.

Has any cultural group solved this practical problem at the present time, even on an intuitive level? One simple fact proves that none has: there is no prosperous population in the world today that has, and has had for some time, a growth rate of zero. Any people that has intuitively identified its optimum point will soon reach it, after which its growth rate becomes and remains zero.

Of course, a positive growth rate might be taken as evidence that a population is below its optimum. However, by any reasonable standards, the most rapidly growing populations on earth today are (in general) the most miserable. This association (which need not be invariable) casts doubt on the optimistic assumption that the positive growth rate of a population is evidence that it has yet to reach its optimum.

We can make little progress in working toward optimum population size until we explicitly exorcize the spirit of Adam Smith in the field of practical demography. In economic affairs, *The Wealth of Nations* (1776) popularized the "invisible hand," the idea that an individual who "intends only his own gain," is, as it were, "led by an invisible hand to promote . . . the public interest."[5] Adam Smith did not assert that this was invariably true, and perhaps neither did any of his followers. But he contributed to a dominant tendency of thought that has ever since interfered with positive action based on rational analysis, namely, the tendency to assume that decisions reached individually will, in fact, be the best decisions for an entire society. If this assumption is correct it justifies the continuance of our present policy of laissez-faire in reproduction. If it is correct we can assume that men will control their individual fecundity so as to produce the optimum population. If the assumption is not correct, we need to reexamine our individual freedoms to see which ones are defensible.

Tragedy of Freedom in a Commons

The rebuttal to the invisible hand in population control is to be found in a scenario first sketched in a little-known pamphlet[6] in 1833 by a mathematical amateur named William Forster Lloyd (1794–1852). We may well call it "the tragedy of the commons," using the word "tragedy" as the philosopher Whitehead used it:[7]

[5] A. Smith, *The Wealth of Nations* (Modern Library, New York, 1937), p. 423.

[6] W. F. Lloyd, *Two Lectures on the Checks to Population* (Oxford Univ. Press, Oxford, England, 1833), reprinted (in part) in *Population, Evolution, and Birth Control*, G. Hardin, Ed. (Freeman, San Francisco, 1964), p. 37.

[7] A. N. Whitehead, *Science and the Modern World* (Mentor, New York, 1948), p. 17.

"The essence of dramatic tragedy is not unhappiness. It resides in the solemnity of the remorseless working of things." He then goes on to say, "This inevitableness of destiny can only be illustrated in terms of human life by incidents which in fact involve unhappiness. For it is only by them that the futility of escape can be made evident in the drama."

The tragedy of the commons develops in this way. Picture a pasture open to all. It is to be expected that each herdsman will try to keep as many cattle as possible on the commons. Such an arrangement may work reasonably satisfactorily for centuries because tribal wars, poaching, and disease keep the numbers of both man and beast well below the carrying capacity of the land. Finally, however, comes the day of reckoning, that is, the day when the long-desired goal of social stability becomes a reality. At this point, the inherent logic of the commons remorselessly generates tragedy.

As a rational being, each herdsman seeks to maximize his gain. Explicitly or implicitly, more or less consciously, he asks, "What is the utility *to me* of adding one more animal to my herd?" This utility has one negative and one positive component.

1. The positive component is a function of the increment of one animal. Since the herdsman receives all the proceeds from the sale of the additional animal, the postive utility is nearly +1.

2. The negative component is a function of the additional overgrazing created by one more animal. Since, however, the effects of overgrazing are shared by all the herdsmen, the negative utility for any particular decision-making herdsman is only a fraction of −1.

Adding together the component partial utilities, the rational herdsman concludes that the only sensible course for him to pursue is to add another animal to his herd. And another; and another. . . . But this is the conclusion reached by each and every rational herdsman sharing a commons. Therein is the tragedy. Each man is locked into a system that compels him to increase his herd without limit—in a world that is limited. Ruin is the destination toward which all men rush, each pursuing his own best interest in a society that believes in the freedom of the commons. Freedom in a commons brings ruin to all.

Some would say that this is a platitude. Would that it were! In a sense, it was learned thousands of years ago, but natural selection favors the forces of psychological denial.[8] The individual benefits as an individual from his ability to deny the truth even though society as a whole, of which he is a part, suffers. Education can counteract the natural tendency to do the wrong thing, but the inexorable succession of generations requires that the basis for this knowledge be constantly refreshed.

A simple incident that occurred a few years ago in Leominster, Massachusetts, shows how perishable the knowledge is. During the Christmas shopping season the parking meters downtown were covered with plastic bags that bore tags reading: "Do not open until after Christmas. Free parking courtesy of the mayor and city council." In other words, facing the prospect of an increased demand for already scarce space, the city fathers reinstituted the system of the commons. (Cynically, we suspect that they gained more votes than they lost by this retrogressive act.)

In an approximate way, the logic of the commons has been understood for a long time, perhaps since the discovery of agriculture or the invention of private property in real estate. But it is understood mostly only in special cases which are not sufficiently generalized. Even at this late date, cattlemen leasing national land on the western ranges demonstrate no more than an ambivalent understanding, in constantly pressuring federal authorities to increase the head count to the point where over-grazing produces erosion and weed-dominance. Likewise, the oceans of the world continue to suffer from the survival of the philosophy of the commons. Maritime nations still respond automatically to the shibboleth of the "freedom of the seas." Professing to believe in the "inexhaustible resources of the oceans," they bring species after species of fish and whales closer to extinction.[9]

The National Parks present another instance of the working out of the tragedy of the commons. At present, they are open to all, without limit. The parks themselves are limited in extent—there is only one Yosemite Valley—whereas population seems to grow

---

[8] G. Hardin, Ed. *Population, Evolution, and Birth Control* (Freeman, San Francisco, 1964), p. 56.

[9] S. McVay, *Sci. Amer.* 216 (No. 8), 13 (1966).

without limit. The values that visitors seek in the parks are
steadily eroded. Plainly, we must soon cease to treat the parks
as commons or they will be of no value to anyone.

What shall we do? We have several options. We might sell them
off as private property. We might keep them as public property,
but allocate the right to enter them. The allocation might be on
the basis of wealth, by the use of an auction system. It might be on
the basis of merit, as defined by some agreed-upon standards. It
might be by lottery. Or it might be on a first-come, first-served
basis, administered to long queues. These, I think, are all the reasonable possibilities. They are all objectionable. But we must
choose—or acquiesce in the destruction of the commons that we
call our National Parks.

Pollution

In a reverse way, the tragedy of the commons reappears in problems of pollution. Here it is not a question of taking something
out of the commons, but of putting something in—sewage, or
chemical, radioactive, and heat wastes into water; noxious and
dangerous fumes into the air; and distracting and unpleasant advertising signs into the line of sight. The calculations of utility
are much the same as before. The rational man finds that his share
of the cost of the wastes he discharges into the commons is less
than the cost of purifying his wastes before releasing them. Since
this is true for everyone, we are locked into a system of "fouling
our own nest," so long as we behave only as independent, rational,
free-enterprisers.

The tragedy of the commons as a food basket is averted by private property, or something formally like it. But the air and waters surrounding us cannot readily be fenced, and so the tragedy of
the commons as a cesspool must be prevented by different means,
by coercive laws or taxing devices that make it cheaper for the pol-
luter to treat his pollutants than to discharge them untreated. We
have not progressed as far with the solution of this problem as we
have with the first. Indeed, our particular concept of private property, which deters us from exhausting the positive resources of the
earth, favors pollution. The owner of a factory on the bank of a
stream—whose property extends to the middle of the stream—often

has difficulty seeing why it is not his natural right to muddy the waters flowing past his door. The law, always behind the times, requires elaborate stitching and fitting to adapt it to this newly perceived aspect of the commons.

The pollution problem is a consequence of population. It did not much matter how a lonely American frontiersman disposed of his waste. "Flowing water purifies itself every 10 miles," my grandfather used to say, and the myth was near enough to the truth when he was a boy, for there were not too many people. But as population became denser, the natural chemical and biological recycling processes became overloaded, calling for a redefinition of property rights.

How To Legislate Temperance?

Analysis of the pollution problem as a function of population density uncovers a not generally recognized principle of morality, namely: *the morality of an act is a function of the state of the system at the time it is performed.*[10] Using the commons as a cesspool does not harm the general public under frontier conditions, because there is no public; the same behavior in a metropolis is unbearable. A hundred and fifty years ago a plainsman could kill an American bison, cut out only the tongue for his dinner, and discard the rest of the animal. He was not in any important sense being wasteful. Today, with only a few thousand bison left, we would be appalled at such behavior.

In passing, it is worth noting that the morality of an act cannot be determined from a photograph. One does not know whether a man killing an elephant or setting fire to the grassland is harming others until one knows the total system in which his act appears. "One picture is worth a thousand words," said an ancient Chinese; but it may take 10,000 words to validate it. It is as tempting to ecologists as it is to reformers in general to try to persuade others by way of the photographic shortcut. But the essence of an argument cannot be photographed: it must be presented rationally—in words.

That morality is system-sensitive escaped the attention of most

10 J. Fletcher, *Situation Ethics* (Westminster, Philadelphia, 1966).

codifiers of ethics in the past. "Thou shalt not . . ." is the form of
traditional ethical directives which make no allowance for particular circumstances. The laws of our society follow the pattern of
ancient ethics, and therefore are poorly suited to governing a complex, crowded, changeable world. Our epicyclic solution is to aug-
ment statutory law with administrative law. Since it is practically
impossible to spell out all the conditions under which it is safe to
burn trash in the back yard or to run an automobile without smog-
control, by law we delegate the details to bureaus. The result is
administrative law, which is rightly feared for an ancient reason—
Quis custodiet ipsos custodes?—"Who shall watch the watchers
themselves?" John Adams said that we must have "a government
of laws and not men." Bureau administrators, trying to evaluate
the morality of acts in the total system, are singularly liable to corruption, producing a government by men, not laws.

Prohibition is easy to legislate (though not necessarily to enforce); but how do we legislate temperance? Experience indicates
that it can be accomplished best through the mediation of administrative law. We limit possibilities unnecessarily if we suppose
that the sentiment of *Quis custodiet* denies us the use of administrative law. We should rather retain the phrase as a perpetual re-
minder of fearful dangers we cannot avoid. The great challenge
facing us now is to invent the corrective feedbacks that are needed
to keep custodians honest. We must find ways to legitimate the
needed authority of both the custodians and the corrective feedbacks.

Freedom To Breed Is Intolerable

The tragedy of the commons is involved in population problems
in another way. In a world governed solely by the principle of
"dog eat dog"—if indeed there ever was such a world—how many
children a family had would not be a matter of public concern.
Parents who bred too exuberantly would leave fewer descendants,
not more, because they would be unable to care adequately for
their children. David Lack and others have found that such a
negative feedback demonstrably controls the fecundity of birds.[11]

11 D. Lack, *The Natural Regulation of Animal Numbers* (Clarendon Press, Oxford, 1954).

But men are not birds, and have not acted like them for millenniums, at least.

If each human family were dependent only on its own resources; *if* the children of improvident parents starved to death; *if,* thus, overbreeding brought its own "punishment" to the germ line—*then* there would be no public interest in controlling the breeding of families. But our society is deeply committed to the welfare state,[12] and hence is confronted with another aspect of the tragedy of the commons.

In a welfare state, how shall we deal with the family, the religion, the race, or the class (or indeed any distinguishable and cohesive group) that adopts overbreeding as a policy to secure its own aggrandizement?[13] To couple the concept of freedom to breed with the belief that everyone born has an equal right to the commons is to lock the world into a tragic course of action.

Unfortunately this is just the course of action that is being pursued by the United Nations. In late 1967, some 30 nations agreed to the following:[14]

> The Universal Declaration of Human Rights describes the family as the natural and fundamental unit of society. It follows that any choice and decision with regard to the size of the family must irrevocably rest with the family itself, and cannot be made by anyone else.

It is painful to have to deny categorically the validity of this right; denying it, one feels as uncomfortable as a resident of Salem, Massachusetts, who denied the reality of witches in the 17th century. At the present time, in liberal quarters, something like a taboo acts to inhibit criticism of the United Nations. There is a feeling that the United Nations is "our last and best hope," that we shouldn't find fault with it; we shouldn't play into the hands of the archconservatives. However, let us not forget what Robert Louis Stevenson said: "The truth that is suppressed by friends is the readiest weapon of the enemy." If we love the truth we must

[12] H. Girvetz, *From Wealth to Welfare* (Stanford Univ. Press, Stanford, Calif., 1950).
[13] G. Hardin, *Perspec. Biol. Med.* 6, 366 (1963).
[14] U. Thant, *Int. Planned Parenthood News,* No. 168 (February 1968), p. 3.

openly deny the validity of the Universal Declaration of Human Rights, even though it is promoted by the United Nations. We should also join with Kingsley Davis[15] in attempting to get Planned Parenthood-World Population to see the error of its ways in embracing the same tragic ideal.

Conscience Is Self-Eliminating

It is a mistake to think that we can control the breeding of mankind in the long run by an appeal to conscience. Charles Galton Darwin made this point when he spoke on the centennial of the publication of his grandfather's great book. The argument is straightforward and Darwinian.

People vary. Confronted with appeals to limit breeding, some people will undoubtedly respond to the plea more than others. Those who have more children will produce a larger fraction of the next generation than those with more susceptible consciences. The difference will be accentuated, generation by generation.

In C. G. Darwin's words: "It may well be that it would take hundreds of generations for the progenitive instinct to develop in this way, but if it should do so, nature would have taken her revenge, and the variety *Homo contracipiens* would become extinct and would be replaced by the variety *Homo progenitivus*."[16]

The argument assumes that conscience or the desire for children (no matter which) is hereditary—but hereditary only in the most general formal sense. The result will be the same whether the attitude is transmitted through germ cells, or exosomatically, to use A. J. Lotka's term. (If one denies the latter possibility as well as the former, then what's the point of education?) The argument has here been stated in the context of the population problem, but it applies equally well to any instance in which society appeals to an individual exploiting a commons to restrain himself for the general good—by means of his conscience. To make such an appeal is to set up a selective system that works toward the elimination of conscience from the race.

15 K. Davis, *Science* 158, 730 (1967).
16 S. Tax, Ed., *Evolution after Darwin* (Univ. of Chicago Press, Chicago, 1960), vol. 2, p. 469.

Pathogenic Effects of Conscience

The long-term disadvantage of an appeal to conscience should be enough to condemn it; but has serious short-term disadvantages as well. If we ask a man who is exploiting a commons to desist "in the name of conscience," what are we saying to him? What does he hear?—not only at the moment but also in the wee small hours of the night when, half asleep, he remembers not merely the words we used but also the nonverbal communication cues we gave him unawares? Sooner or later, consciously or subconsciously, he senses that he has received two communications, and that they are contradictory: (i) (intended communication) "If you don't do as we ask, we will openly condemn you for not acting like a responsible citizen"; (ii) (the unintended communication) "If you *do* behave as we ask, we will secretly condemn you for a simpleton who can be shamed into standing aside while the rest of us exploit the commons."

Everyman then is caught in what Bateson has called a "double bind." Bateson and his co-workers have made a plausible case for viewing the double bind as an important causative factor in the genesis of schizophrenia.[17] The double bind may not always be so damaging, but it always endangers the mental health of anyone to whom it is applied. "A bad conscience," said Nietzsche, "is a kind of illness."

To conjure up a conscience in others is tempting to anyone who wishes to extend his control beyond the legal limits. Leaders at the highest level succumb to this temptation. Has any President during the past generation failed to call on labor unions to moderate voluntarily their demands for higher wages, or to steel companies to honor voluntary guidelines on prices? I can recall none. The rhetoric used on such occasions is designed to produce feelings of guilt in noncooperators.

For centuries it was assumed without proof that guilt was a valuable, perhaps even an indispensable, ingredient of the civilized life. Now, in this post-Freudian world, we doubt it.

Paul Goodman speaks from the modern point of view when he

---

[17] G. Bateson, D. D. Jackson, J. Haley, J. Weakland, *Behav. Sci.* 1, 251 (1956).

says: "No good has ever come from feeling guilty, neither intelligence, policy, nor compassion. The guilty do not pay attention to the object but only to themselves, and not even to their own interests, which might make sense, but to their anxieties."[18]

One does not have to be a professional psychiatrist to see the consequences of anxiety. We in the Western world are just emerging from a dreadful two-centuries-long Dark Ages of Eros that was sustained partly by prohibition laws, but perhaps more effectively by the anxiety-generating mechanisms of education. Alex Comfort has told the story well in *The Anxiety Makers;*[19] it is not a pretty one.

Since proof is difficult, we may even concede that the results of anxiety may sometimes, from certain points of view, be desirable. The larger question we should ask is whether, as a matter of policy, we should ever encourage the use of a technique the tendency (if not the intention) of which is psychologically pathogenic. We hear much talk these days of responsible parenthood; the coupled words are incorporated into the titles of some organizations devoted to birth control. Some people have proposed massive propaganda campaigns to instill responsibility into the nation's (or the world's) breeders. But what is the meaning of the word responsibility in this context? Is it not merely a synonym for the word conscience? When we use the word responsibility in the absence of substantial sanctions are we not trying to browbeat a free man in a commons into acting against his own interest? Responsibility is a verbal counterfeit for a substantial *quid pro quo*. It is an attempt to get something for nothing.

If the word responsibility is to be used at all, I suggest that it be in the sense Charles Frankel uses it.[20] "Responsibility," says this philosopher, "is the product of definite social arrangements." Notice that Frankel calls for social arrangements—not propaganda.

Mutual Coercion Mutually Agreed Upon

The social arrangements that produce responsibility are arrangements that create coercion, of some sort. Consider bank-robbing.

[18] P. Goodman, *New York Rev. Books* 10(8), 22 (23 May 1968).
[19] A. Comfort, *The Anxiety Makers* (Nelson, London, 1967).
[20] C. Frankel, *The Case for Modern Man* (Harper, New York, 1955), p. 203.

The man who takes money from a bank acts as if the bank were a commons. How do we prevent such action? Certainly not by trying to control his behavior solely by a verbal appeal to his sense of responsibility. Rather than rely on propaganda we follow Frankel's lead and insist that a bank is not a commons; we seek the definite social arrangements that will keep it from becoming a commons. That we thereby infringe on the freedom of would-be robbers we neither deny nor regret.

The morality of bank-robbing is particularly easy to understand because we accept complete prohibition of this activity. We are willing to say "Thou shalt not rob banks," without providing for exceptions. But temperance also can be created by coercion. Taxing is a good coercive device. To keep downtown shoppers temperate in their use of parking space we introduce parking meters for short periods, and traffic fines for longer ones. We need not actually forbid a citizen to park as long as he wants to; we need merely make it increasingly expensive for him to do so. Not prohibition, but carefully biased options are what we offer him. A Madison Avenue man might call this persuasion; I prefer the greater candor of the word coercion.

Coercion is a dirty word to most liberals now, but it need not forever be so. As with the four-letter words, its dirtiness can be cleansed away by exposure to the light, by saying it over and over without apology or embarrassment. To many, the word coercion implies arbitrary decisions of distant and irresponsible bureaucrats; but this is not a necessary part of its meaning. The only kind of coercion I recommend is mutual coercion, mutually agreed upon by the majority of the people affected.

To say that we mutually agree to coercion is not to say that we are required to enjoy it, or even to pretend we enjoy it. Who enjoys taxes? We all grumble about them. But we accept compulsory taxes because we recognize that voluntary taxes would favor the conscienceless. We institute and (grumblingly) support taxes and other coercive devices to escape the horror of the commons.

An alternative to the commons need not be perfectly just to be preferable. With real estate and other material goods, the alternative we have chosen is the institution of private property coupled with legal inheritance. Is this system perfectly just? As a genetically trained biologist I deny that it is. It seems to me that, if there are to be differences in individual inheritance, legal possession should be perfectly correlated with biological inheritance—that those who are biologically more fit to be the custodians of property and power should legally inherit more. But genetic recombination continually makes a mockery of the doctrine of "like father, like son" implicit in our laws of legal inheritance. An idiot can inherit millions, and a trust fund can keep his estate intact. We must admit that our legal system of private property plus inheritance is unjust—but we put up with it because we are not convinced, at the moment, that anyone has invented a better system. The alternative of the commons is too horrifying to contemplate. Injustice is preferable to total ruin.

It is one of the peculiarities of the warfare between reform and the status quo that it is thoughtlessly governed by a double standard. Whenever a reform measure is proposed it is often defeated when its opponents triumphantly discover a flaw in it. As Kingsley Davis has pointed out,[21] worshippers of the status quo sometimes imply that no reform is possible without unanimous agreement, an implication contrary to historical fact. As nearly as I can make out, automatic rejection of proposed reforms is based on one of two unconscious assumptions: (i) that the status quo is perfect; or (ii) that the choice we face is between reform and no action; if the proposed reform is imperfect, we presumably should take no action at all, while we wait for a perfect proposal.

But we can never do nothing. That which we have done for thousands of years is also action. It also produces evils. Once we are aware that the status quo is action, we can then compare its discoverable advantages and disadvantages with the predicted advantages and disadvantages of the proposed reform, discounting as best we can for our lack of experience. On the basis of such a comparison, we can make a rational decision which will not involve the unworkable assumption that only perfect systems are tolerable.

21 J. D. Roslansky, *Genetics and the Future of Man* (Appleton-Century-Crofts, New York, 1966), p. 177.

Recognition of Necessity

Perhaps the simplest summary of this analysis of man's population problems is this: the commons, if justifiable at all, is justifiable only under conditions of low-population density. As the human population has increased, the commons has had to be abandoned in one aspect after another.

First we abandoned the commons in food gathering, enclosing farm land and restricting pastures and hunting and fishing areas. These restrictions are still not complete throughout the world.

Somewhat later we saw that the commons as a place for waste disposal would also have to be abandoned. Restrictions on the disposal of domestic sewage are widely accepted in the Western world; we are still struggling to close the commons to pollution by automobiles, factories, insecticide sprayers, fertilizing operations, and atomic energy installations.

In a still more embryonic state is our recognition of the evils of the commons in matters of pleasure. There is almost no restriction on the propagation of sound waves in the public medium. The shopping public is assaulted with mindless music, without its consent. Our government is paying out billions of dollars to create supersonic transport which will disturb 50,000 people for every one person who is whisked from coast to coast 3 hours faster. Advertisers muddy the airwaves of radio and television and pollute the view of travelers. We are a long way from outlawing the commons in matters of pleasure. Is this because our Puritan inheritance makes us view pleasure as something of a sin, and pain (that is, the pollution of advertising) as the sign of virtue?

Every new enclosure of the commons involves the infringement of somebody's personal liberty. Infringements made in the distant past are accepted because no contemporary complains of a loss. It is the newly proposed infringements that we vigorously oppose; cries of "rights" and "freedom" fill the air. But what does "freedom" mean? When men mutually agreed to pass laws against robbing, mankind became more free, not less so. Individuals locked into the logic of the commons are free only to bring on universal ruin; once they see the necessity of mutual coercion, they become free to pursue other goals. I believe it was Hegel who said, "Freedom is the recognition of necessity."

The most important aspect of necessity that we must now recognize, is the necessity of abandoning the commons in breeding. No technical solution can rescue us from the misery of overpopulation. Freedom to breed will bring ruin to all. At the moment, to avoid hard decisions many of us are tempted to propagandize for conscience and responsible parenthood. The temptation must be resisted, because an appeal to independently acting consciences selects for the disappearance of all conscience in the long run, and an increase in anxiety in the short.

The only way we can preserve and nurture other and more precious freedoms is by relinquishing the freedom to breed, and that very soon. "Freedom is the recognition of necessity"—and it is the role of education to reveal to all the necessity of abandoning the freedom to breed. Only so, can we put an end to this aspect of the tragedy of the commons.

QUESTIONS AND SUGGESTIONS

1. What are the reasons for the order of presentation of the various parts of this article?

2. Hardin's position is still a controversial one. What arguments might be advanced against his?

3. Write a review article on the population issue as it stands now. If you wish, choose a more limited topic for a review article, a form that surveys the current state of thinking on a particular topic.

4. Compare Carson's tone to Hardin's. What role does persuasion play in writing about science?

LYNN MARGULIS
AND DORION SAGAN

Threats to DNA and the Emergence of Sexuality

LYNN MARGULIS (b. 1938) is University Professor of Biology at Boston University. She is a member of the National Academy of Sciences and a fellow of the American Association for the Advancement of Science. Margulis, who also writes book reviews, is the author of *The Five Kingdoms* with Karlene Schwartz (1987, second edition), and two books with DORION SAGAN (b. 1959), *The Garden of Microbial Delights* (1988) and *Biospheres from Earth to Space* (1989). Her textbook *Handbook of Protoctista* is soon to be accompanied by *The Handbook of Protoctista Glossary* (1990).

Time and the Early Earth

Our narrative history of sexuality from now on will follow, insofar as possible, the chronology of life on Earth. Cosmologists, nuclear physicists, astronomers, and space scientists have colossally changed many of the most basic human beliefs. Working independently, they have produced myriads of diverse data for investigative minds to sort out and integrate. We slowly build a picture of the timescape (Calder, 1983). In the most prevalent model, the universe, forming in the biggest bank imaginable, came into being in about three minutes 13,500 million years ago (Weinberg, 1977), and it has been expanding ever since. In about a second after it was born, matter from that bang had traveled outward three light-years. Three minutes later this matter—particles but not yet atoms—heated to a billion degrees centigrade, had covered some forty light-years.

The universe is primarily made up of hydrogen and helium, the first two elements of the periodic table. The heavier elements making up the Sun and planets came later. Most were produced in the rare celestial events called "supernovae." Supernovae are extremely bright but short-lived explosions that occur as stars or stellar clouds violently collapse into neutron stars and release most of their mass in the form of energy. Atoms such as those that eventually found their way into the bodies of autopoietic beings began as products of these spectacular explosions. Such matter—spinning, expanding, exploding—became localized in many parts of the universe, including our solar system. Some cosmologists feel that not only can the origin of the Sun and its planets now be explained, but a chronology of the life history of the solar system, starting just this side of 5,000 million years ago, can be drawn up.

Some 4,650 million years ago a presolar cloud of gas and dust hovered in our region of the universe. Shock waves from the spiral arms of passing galaxies swept through the cloud that was to become the Sun and created a huge new star. The destiny of stars depends on their total mass: this ill-fated star giant did not last, but it provided matter for future bodies. Some 4,550 million years ago a spiral arm passed by again and made still another star. The shock wave from this supernova, the cosmogonists tell us, was enough to induce the precipitate collapse of dust and gas that converted the presolar cloud into our star, the Sun. The same series of events that made the Sun created its entourage of planets, including Earth. The Moon was apparently captured by the same gravitational forces that accreted Earth into a solid body some 4,500 million years ago. About 50 million years after that (4,500–4,450 million years ago) the concentric levels of Earth differentiated: molten iron and nickel entered the core, leaving toward the exterior the lighter elements that make up the surface crust. Energy was provided to the mix from continuous cosmic bombardments and radioactive decay from constituent elements. Well differentiated and fully made, Earth and Moon settled down about 4,450 million years ago. The meteoric and planetoid bombardment did not stop abruptly, however. We know from the cratered terrain of the Moon and inner planets that, for several thousand years at least, though probably with lessening intensity, the surface of Earth was battered and bruised by extraterrestrial impacts from meteorites of many sizes composed of a variety of cosmic materials.

The *Explorer 10* satellite has recently returned data about the output of radiation from Sun-like stars, permitting us to reconstruct the probable output of light energy of the early Sun (Canuto et al., 1982). Output was so great that, if it had remained unattenuated by the atmosphere (an atmosphere with a composition that seems most reasonable to astronomers and geologists for the early Earth), life should not have evolved at all. Deadly ultraviolet light, including those particular wavelengths absorbed by DNA, RNA, and proteins, shone down mercilessly. If, as is probable, the early atmosphere was composed of nitrogen, water vapor, and carbon dioxide, life was challenged from the time of its origin with the "danger/opportunity" of dealing with the large ultraviolet fluxes. As was often to happen later in the evolutionary story, under different circumstances, life-threatening danger became transformed into life-producing opportunity. Microbes had to deal with the large fluxes of ultraviolet radiation that pierced the early atmosphere.

Today, free oxygen forms ozone and ozone very effectively shields us from ultraviolet light of these threatening wavelengths (Fig. 1). But planetary oxygen, scientists now agree, has emanated as a waste product from later life (Cloud, 1983). Most recent estimates of the maximum amount of oxygen that could have been around in the atmosphere prior to the onset of oxygen production by life are minuscule indeed, about one part in 100 million by volume (Canuto et al., 1982). Thus from the beginning of life ultraviolet and visible light threatened the very integrity of life, as will be explained below.

First Life and Multicellularity

The scientific community, as well as members of the public, have been astonished during the last two decades to learn the probabilities of life on the planet. We now realize that there is direct evidence that life has existed nearly as long as there has been a planet Earth. The oldest dated rocks, rocks that have never remelted since their formation, are about 3,900 million years old. They may contain evidence for life on Earth at that time—according to optimists. Unfortunately, these oldest rocks, of the Archean Eon (see Table 1) have been too much heated and pressurized for fossils of living beings to have survived. They do, however, contain large quantities of carbonaceous matter. In such rocks from Greenland and Labrador, carbon in the form of graphite, with tiny quantities of more complex organic compounds, may represent traces of the early events that led to life on this planet. Yet the evidence here is hardly conclusive.

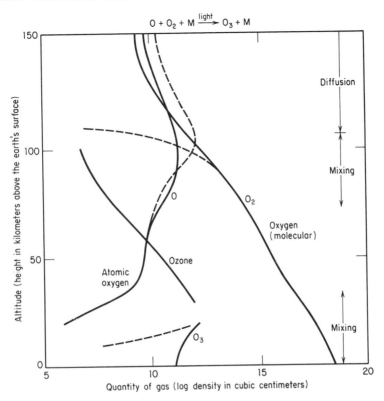

$$O + O_2 + M \xrightarrow{\text{light}} O_3 + M$$

Figure 1.   Ozone in the Earth's atmosphere. The photochemical reaction producing ozone from oxygen is shown at the top; M = any molecule entering the reaction. Solid lines are actual quantities of the gases, dotted lines indicate calculated values based on the assumption of photochemical equilibrium. Mixing and diffusion at the altitudes in the atmosphere indicated at the right cause departure of the values from equilibrium. For details see Goody and Walker, 1972, and Walker, 1986. (Drawing by Christie Lyons.)

A little closer to the present, though, the evidence for early life becomes impressive. That bacteria of some kind inhabited Earth as far back as 3,500 million years ago is supported by a rather rich series of observations, growing richer all the time. From both the Warrawoona Formation, at North Pole in western Australia, and the Fig Tree Formation of sedimentary rocks, in the Swaziland

Table 1. Geological Time*

Eon	Era	Period	Epoch	Age in Millions of Years
PRE-PHANEROZOIC			*(Origin of Earth)*	4,500
HADEAN				
ARCHEAN			*(Oldest rocks)*	3,900
				2,600
PROTEROZOIC	Aphebian / Riphean			2,000
				1,000
	Vendian			580
PHANEROZOIC	Paleozoic	CAMBRIAN		500
		ORDOVICIAN		440
		SILURIAN		400
		DEVONIAN		345
		CARBONIFEROUS		290
		PERMIAN		245
	Mesozoic	TRIASSIC		195
		JURASSIC		138
		CRETACEOUS		66
	Cenozoic	PALEOGENE	Paleocene	54
			Eocene	38
		TERTIARY	Oligocene	26
			Miocene	7
		NEOGENE	Pliocene	2
		QUATERNA	Pleistocene	0.01
			Recent	0

* Not to scale.

System from southern Africa, a concatenation of evidence convinces us of the antiquity of life. In the late 1960s and early 1970s Elso Barghoorn, professor of biology and geology at Harvard University, and his colleagues came forth with impressive evidence that pushed back the fossil record about a billion years (Barghoorn, 1971).

Few now doubt that these fossils were authentic remains of living organisms or that the early life Barghoorn discovered was bacterial. From the beginning it was clear that spheroidal unicells were present, some apparently in the process of division. Later filamentous (and therefore multicellular) bacteria were discovered both in the same material and in rocks of comparable age from Warrawoona, Australia. It is not surprising that multicellular life follows on the heels of unicellular forms, because offspring cells so often fail to separate from parent cells after division. Strings of cells (called "pseudofilaments" because they have no cell connections) form first, then true filaments evolve. Probably because of this failure of quick separation after reproduction, the first and simplest sort of multicellularity arose many times in many bacterial lineages. Multicellularity, a characteristic of many species, is another aspect of life that is not restricted to sexual organisms, to animals and plants, or even to eukaryotes. Indeed, the term *multicellular being* does not necessarily designate an animal or plant (Margulis, Mehos, and Kraeski, 1983). Furthermore, multicellularity in bacteria may be accompanied with some differentiation in the sense that the cells of the multicellular organisms differ from each other both morphologically and physiologically (see Table 2).

The story of the origin of life on an early, turbulent Earth is being told in growing richness of detail (Day, 1984; Miller and Orgel, 1974). The question we are asking is how sexuality began within the earliest life. How, for example, did the Archean ultraviolet light flux affect sexuality?

The Dangerous Opportunity of Ultraviolet Light

Sunlight, unattenuated by ozone, posed an incessant threat to DNA integrity. Ultraviolet light at first was not separated from the visible light that powered early living systems. Yet within a

Table 2.   Bacterial Differentiation

Type of Organism	Examples	Comments
Endospores	Clostridium, Arthromitus, Bacillus	Extensive developmental cycle showing formation of new membranes
Prosthecae (stalk formation)	Caulobacter and other prosthecate bacteria	May alternate with flagellated stage
Heterocysts	Cyanobacteria	Nitrogen fixation
Thallus	Cyanobacteria, purple photosynthetic bacteria, zoogloeas	Differentiated basal cells; cubes may be formed or flat sheets of cells may cover extensive areas
Exospores	Actinobacteria	Borne on tips of trichomes
Colonial structures	Arthrobacter, some pseudomonads and bacilli	Complex fruiting structures in myxobacteria group, exudates from cells
Akinetes	Cyanobacteria	Thick-walled dissemination structures

geologically short time, perhaps a few hundred million years, all life on Earth became dependent, as it still is today, on photosynthesis. Staying in the sunlight was, and still is, obligate for photosynthetic bacteria, and hence for all early life. To absorb the necessary and less damaging visible light and avoid the life-threatening ultraviolet light was a problem facing all light-requiring organisms before the formation of the ozone shield.

The conservatism of living systems permits us to reconstruct their past. The first clue to the origin of sexual systems came from an observation that is well known but almost never discussed in the evolutionary literature: mutations leading to the loss of sexual ability in the colon bacteria simultaneously render these bacteria extremely sensitive to ultraviolet light. Likewise, mutations leading to the loss of the ability to cope with ultraviolet light may simultaneously lead to the total destruction of the genetic recombination system. *Escherichia coli* bacteria no longer able to undergo recombination are called "rec minus" mutants. These mutants, in many cases, are hundreds of times more prone to death by ultraviolet light.

A few of these cases are understood in detail. The major effect of ultraviolet light on DNA is the formation of lethal knots in the linear sequence where two thymine bases are found next to each other. Four-membered carbon rings, called "thymine dimers," are produced—as if in the steps of the ladder the bases had twisted and attached to their neighbors above and below rather than to those on the opposite side, as they should. Thymine dimers knot up DNA so badly that replication and information transfer to RNA are interfered with and death ensues.

The inevitability of ultraviolet-induced formation of thymine dimers and subsequent death must have been averted early on in the history of life on this planet. Even today there are elaborate and complex methods of enzymatically overcoming the ultraviolet death threat. . . . A remarkable variety of mechanisms ensuring the retention of the integrity of DNA molecules exists in cells. Three classes of these have been recognized: (1) fidelity mechanisms associated with DNA replication (for example, biochemistry involving proofreading and mismatch correction processes), (2) detoxification mechanisms (primarily against oxygen), and (3) many kinds of repair (Haynes, 1985).

In retrospect, the arguments for ultraviolet radiation leading to the first sex among bacteria is simple. DNA, RNA, and protein strongly absorb ultraviolet light. The early atmosphere of Earth was composed primarily of water vapor, nitrogen, and carbon dioxide, none of which absorbs ultraviolet light in the part of the spectrum that organic macromolecules absorb (260–280 nanometers). Thus all microbes living in the open before the appearance of the ozone layer (which absorbs and therefore protects against ultraviolet light) must have been subjected to threats to the integrity of their macromolecules.

Particularly threatened were photosynthetic bacteria that required visible light. An entire battery of mechanisms for ultraviolet protection evolved, including enzymes that directly repair damage to DNA integrity. At least some of these enzymes act by removing damaged DNA sequences (such as thymine dimers) and resynthesizing DNA, using available intact DNA as a complement.

Insofar as the damaged DNA uses its complement to guide enzymatic repair, the system is merely a repair system. If no complement is available in the cell itself the DNA is irreparable and the cells die. If DNA from any other cell source is used as a complement—and that source may be any small replicon, such as a bacteriophage, or it may be appropriate DNA in solution (called "transforming DNA")—the repair process is a two-parent one and, by definition, is a form of sexuality. The splicing and polymerase enzymes for repair became the enzymes of sexuality; DNA recombination in nature is far older than it is in the laboratory. Thus, in the evolutionary sense *ultraviolet repair preadapted bacteria to sexuality.* The source of the complementary DNA in some cases was from an entirely different cell with its own different, but homologous, DNA. When the source of the new DNA was different from that of the damaged DNA, the repair process became a form of prokaryotic sexuality: new DNA was formed from more than a single parental source.

Later, after the appearance of some atmospheric oxygen and with it the ozone screen, ultraviolet light became a less serious threat. Yet ultraviolet repair systems were still retained in many organisms because they had become part of sexual and other systems that by now served various functions in addition to ultraviolet protection. These DNA repair systems were retained for different reasons in subsequent lineages.

We conclude that the first sort of microbial sexuality was a direct response to life-threatening danger. Apparently this scenario, whereby the preadaptation to sexuality evolved as a method of survival, applies only to a subset of the members of the Monera kingdom. For example, gram-positive bacteria, including all the endospore-forming organisms, seldom pair or form direct cell connections, although some gram-positive bacteria have small replicon—mediated recombination (Clewell, 1985). Though cells may come together in clumps, sexuality in these organisms is much less well understood than that in gram-negative bacteria. No gram-positive bacteria are photosynthetic. Perhaps, having no need for sunlight, they were secluded from the zones of ultraviolet radiation that applied fierce and constant selection pressure toward the refinement of DNA repair mechanisms. The atmospheric build-up of oxygen and ozone about a billion years ago was due to the spread of active, water-using photosynthesis. After the formation of an atmosphere that shielded the planet against ultraviolet radiation, the inducement to evolve such mechanisms must have been lessened considerably.

The Imperative of Light and the Double Jeopardy of Photosynthesizers

Some form of resistance to ultraviolet light in the early Archean Eon was imperative and many methods that could have been used are known. The idea that resistance to ultraviolet light is an ancient legacy is supported by the observation that obligate anaerobes, organisms such as *Clostridium,* are poisoned by oxygen. The fact that organisms tend to retain their most important attributes suggest an unbroken lineage, a continuity between ancient and modern forms. The ancestors of *Clostridium* must have evolved prior to the entry of oxygen into the atmosphere, for they are far more resistant to ultraviolet light than are aerobes. Furthermore, when microbes that can be grown under either anaerobic or aerobic conditions are tested for resistance to ultraviolet radiation delivered anaerobically, the same organism is more resistant if grown afterward anaerobically than if grown in the presence of oxygen (Rambler, 1980; Rambler and Margulis, 1980). One assumes that resistance of anaerobes to ultraviolet light is well developed because these organisms evolved when such radiation was a major threat to their existence. Resistance may be maintained because the entire ultraviolet-response system has found other uses. This, however, is not altogether clear.

Recombination, presumably, is only one of many methods anaerobic microbes first developed to counter the ultraviolet threat. Another method is spore production. Spores, bacterial structures resistant to heat and desiccation, are much more hardy in the presence of threatening radiation than are the corresponding growing forms, or "vegetative structures," of the spore-forming bacteria. Spores can delay germination until sundown or until they are covered with protective layers of water, other bacteria, or organic scums. Another simple solution to the ultraviolet threat is the "sunglasses" ploy. Staying immersed in water-soluble compounds or covered by insoluble compounds that absorb ultraviolet radiation is an example of this method. Bacteria covered by the remains of other live bacteria or even bacterial debris are protected from the ravages of ultraviolet light, whereas unshielded bacteria under the same conditions die. Cyanobacteria may employ the method of living in the presence of high concentrations of nitrate or nitrite (Rambler, 1980). Some bacteria evidently protect themselves by appropriate choice of medium. Others, however, activate repair enzymes specifically for the purpose of healing the damage. At least one class of these enzymes is stimulated by ordinary visible light. If a microbe is placed in the dark after ultraviolet treatment it will die, whereas if it is placed in the light it will live. This survival in light is the result of photoreactivation, that is, repair of DNA by light-dependent enzymes. Since in nature visible light accompanies the ultraviolet, the bacteria must have availed themselves of photoreactivity from the very beginning.

Cyanobacteria are oxygen-producing, gram-negative, photosynthetic bacteria. Widely distributed and tenacious microbes, they have apparently such a large bag of tricks to protect themselves against the constant annoyance of ultraviolet light (see Table 3) that they—or at least the well-studied laboratory weeds—never developed sex at all. Their relatives, photosynthetic bacteria that do not produce oxygen (such as *Rhodopseudomonas*), apparently do engage in *E. coli*–style sexuality (Marrs, 1983).

Table 3.   Bacterial Ultraviolet Repair Systems

Photoreactivation of ultraviolet damage by visible light

Dark repair: delay of growth until enzymes restore thymines from thymine dimers

Excision of damaged bases: excision repair, damaged bases spliced out

Recombinational repair: resynthesis of DNA molecule from undamaged fragments as template

SOS repair: DNA chain growth across damaged segments (an error-prone process)

Note: All but the first, which requires exposure to visible light, are dark-repair processes.

Ultraviolet-Induced Viral Dissemination

The relationship between prokaryotic sexuality and ultraviolet light has been known and used as a practical tool since D'Herelle's discovery of "filterable agents," more commonly known as viruses, early in this century (D'Herelle 1926). Although the fact is usually not mentioned in an evolutionary context it is well known that mild ultraviolet-light treatment, such as the placement of an appropriate culture of bacteria beneath a germicidal mercury lamp ultraviolet source for less than a minute, induces the emergence from the cells of various kinds of small genetic entities. Death of the bacterial cell accompanies the release of entities called "bacteriophages." The ultraviolet-treated cell lyses, that is, bursts open. The best-studied example of this release is that of the lysogenic bacteriophages. These viruses "live" inside their bacterial hosts. Integrated within the hosts' genetic material, they reproduce whenever the host does.

Under the proper conditions the latent phages become active and destroy their hosts. The entire activity, known as "lysogeny," or "phage burst," is routinely induced by placing the lysogenic host cells beneath an ultraviolet lamp. If we regard viruses as part of a legacy of ultraviolet-induced cell destruction, it should come as no surprise that even today ultraviolet irradiation leads to the emergence of bits of genome. Inside a phage particle (virion) the dormant DNA is safe within a protein case. It is reasonable to assume, then, that the phage particle, often with a few bacterial genes strung onto it, has a better chance of ultimately reaching a safer realm than do naked bits of genome exposed directly to irradiation.

Not only bacterial viruses but also animal and plant viruses, as well as various plasmids (which may be thought of as viruses without the usual protein coats), will often burst out of cells that are exposed to ultraviolet light. Dense genetic particles with protein coats, carrying a portion of the genome of the autopoietic host, are protected from thymine-dimer production. Safe from lethal light, the genes are available for eventual penetration and integration into the linear order of a second autopoietic host.

Intrinsic Sexuality of Viruses

Often viruses and other small genetic entities carry with them bits of informational DNA that they have picked up from their former hosts. This DNA may code for one or another useful trait. Not being autopoietic, these entities do not survive unless they enter an autopoietic host. When they infect such new hosts, as is well known, they may combine their DNA with that of their new host. More specifically, the DNA of the viruses and plasmids, by the action of a battery of appropriate enzymes, becomes integrated into the linear genetic order of the chromoneme of the bacterial cell. The release by one bacterium and subsequent uptake of viruses by another probably evolved as a response to ultraviolet light and to other dire environmental conditions as well. Since the fundamental process of viral transfer between hosts also involves the formation of a new piece of DNA from more than a single parental source, it is like DNA repair of ultraviolet damage, a form of chromonemal sexuality.

Sexuality of viruses has been known as long as viruses themselves have been studied. A clue to this intrinsic sexuality was observed upon mixing a virus of one sort (one, say, that causes small, ruffled plaques in its bacterial host when it bursts out) with that of another (one causing large, smooth plaques upon bursting). The mixture of viruses produced not only copies of the parental types but also "recombinant" types (ruffled, large types and smooth, small ones). The reason for the recombination is the tendency of viral DNA inside a host to recombine in such a way that the viruses coming out carry DNA from different sources. The rules governing the proportions of parental and recombinant forms of viruses are neither universal nor simple. They can change with conditions that affect the chemistry of DNA. Suffice it to say that if the conditions for subsequent life are permissive (that is, if all recombinants— smooth, rough, large, and small—can survive) recombinants *do* survive. Sex, in the sense of viral recombination, was not necessary for reproduction or anything else; it simply occurred in the course of survival. The enzymes to break, patch, repair, and recombine DNA were present and functioning.

We have indicated that prokaryotic sexuality arose as a response

LYNN MARGULIS AND DORION SAGAN   317

to the threat of ultraviolet-forced disintegration of nucleic acids.
It is likely that an analogous tale can be told of responses to chemical threats as well. These include threats of toxic concentrations
of mercury or manganese, the resistance to which is known to be
borne on the sort of small replicons called plasmids." But, because
the detailed organic and metallic chemistry of the early Archean
Eon is more difficult to reconstruct than is radiation quality and
flux, we have confined ourselves to the example of ultraviolet light.

In summary, then, prokaryotic sexuality—recombination on the
DNA level—is best understood as a possible response to ultraviolet
irradiation and other threats to DNA survival. These survival
mechanisms involved the borrowing of an undamaged DNA. To
be usable as a complement from which a good DNA copy could
then be made, the second DNA had to be recognizably similar to
the first. In this step alone we see the appearance of at least a
second parent.

Prokaryotic sexuality, first an enzymatic response to ultraviolet
or chemical threats to the linear integrity of DNA, later became
co-opted for other tasks. An example of such later co-opting can
still be seen today. The distribution throughout the environment
of certain types of plasmids and viruses is revealing. It has been
observed (Silver, 1983; Rosson and Nealson, 1982a, 1982b), Lid-
strom, Engebrecht, and Nealson, 1983) that, in environments bearing toxic quantities of metallic compounds and organic poisons
(including antibiotics), the genetic factors carrying resistance to
these insults are borne on viruses, plasmids, or other small, mobile
replicons. These entities quickly reproduce and are passed from
vulnerable cell to cell in direct response to the environmental
toxin. (For a review of genetic transfer by such transient entities
in prokaryotes, see Sonea and Panisset, 1983.) Recombination of
microbial DNA, a legacy of response to traumatic ultraviolet radiation in the Archean Eon, is the first step in the long and winding
sexual pageant of planetary life.

QUESTIONS AND SUGGESTIONS

1. What is the historical relationship between bacteria and oxygen?
2. Why is sunlight a threat to DNA?

3. Describe the process by which bacteria is understood to have a sex.

4. What role do the tables play in Margulis's essay?

5. Identify all the essays dealing with DNA in this reader. Compile an annotated bibliography on the progress of DNA research as reflected here. Conduct some independent research into topics pertaining to the history of DNA and comment upon the representative nature of the essays presented here.

GEORGE AND MURIEL BEADLE

The Mendelian Laws

GEORGE BEADLE (1903–1989) helped put modern genetics on a chemical basis by demonstrating the single-gene single-enzyme hypothesis, for which he shared the 1958 Nobel Prize in medicine or physiology. Beginning in 1941, working with *Neurospora crassa*, he and E. L. Tatum produced mutations using X-rays and fed 1000 mutated, sprouted, but non-growing spore cultures special diets, noting which promoted growth and which did not. The 299th culture showed that vitamin B_6, pyridoxine, restored the strain to proper growth. Beadle and Tatum were able to conclude that radiation had damaged a specific gene responsible for producing an enzyme essential for synthesis of vitamin B_6. Their work has had great value recently in the control by chemical treatment of genetic disease. Beadle led a distinguished academic career, serving as President of the University of Chicago from 1961 to 1968 and retaining a teaching post until 1975. He is the author of *An Introduction to Genetics*, with A. H. Sturtevant (1939), and *Genetics and Modern Biology* (1963). MURIEL BEADLE (b. 1915) collaborated with him in writing *The Language of Life*, from which "The Mendelian Laws" is taken. She did the writing; he supplied the content.

The Beadles said that their audience is not so much the non-scientist as the person whose instruction in science occurred since the mid-1950s, when rapid progress in genetics took place. The purpose of this essay is to supply the background in classical genetics preliminary to an exposition of twentieth-century developments in the field.

You've seen him in the movies: the scientist who is infallible, insensitive, and coldly objective—little more than an animated computer in a white lab coat. He takes measurements and records results as if the collection of data were his sole object in life; and if a meaningful pattern emerges it comes as a blinding surprise. The assumption is that if one gathers enough facts about something, the relationships between those facts will spontaneously reveal themselves.

Nonsense.

In the real world of science, the investigator almost always knows what he's looking for before he starts. His observations are usually undertaken to prove the validity of an idea, and his emotions are as deeply engaged as those of a businessman planning a sales campaign, a general mapping out strategy, or a hunter stalking big game.

It's true that scientists strive for objectivity; what's more, they achieve it more often than other men. But they are no more capable than other men of maintaining absolute neutrality toward the outcome of their work. Nor could *you*, if you were testing an original hypothesis that you believed to be both unique and imaginative. Who among us would not like to make a successful thrust into the unknown, to find a missing link, to break a code?

Scientists are more curious than most of their fellows, more intelligent than many, and the best of them are as creative as the best composers, poets, or painters. But they are equally human. Thus they are liable to error, subject to luck, and affected by the political or emotional climate of their times in much the same way as anyone else is.

The myth of infallibility evaporates when one thinks of the number of great ideas in science whose originators were correct only in general but wrong in detail. Dalton, for example, gets credit for the atomic theory as we know it today—yet his formulas for figuring atomic weights were basically incorrect. Copernicus was mistaken in the particulars of his sun-centered universe; it doesn't explain the movements of the planets any better than Ptolemy's earth-centered universe did. Newton amended Kepler; and even Newton's laws of physics have been modified (although not in ways important to the layman) by Einstein.

It may be no easier for a scientist to challenge the prevailing thought of his time than for any other man. Witness Darwin's excessive caution in avoiding mention of *human* evolution when he wrote *Origin of Species*. He realized the implications of his work, and anticipated the storm of public protest.

It's even harder for a scientist to defend an opinion that is unpopular among other scientists. The Swedish chemist Arrhenius, for example, was almost denied his Ph.D. because of wild ideas expressed in his thesis about the existence of particles he called "ions"; and although there is a Cinderella twist to his story—nineteen years later, after electrons had been discovered, he was awarded a Nobel prize for the same research that had nearly lost him his degree—there are numerous examples of other scientists who went to their graves with *their* wild ideas unnoticed or unvalidated. One such was the English physician Garrod, the forgotten man of biochemistry. He suggested that genes control chemical reactions by the use of enzymes, but his theories fitted neither into the context of scientific thought of his time nor into an established scientific discipline, and neither chemists nor biologists took proper note of them.

Luck, too, has played as much of a role in scientific discovery as in any other human endeavor. The German astronomer Kepler, for example, made two mistakes in simple arithmetic in calculating the orbit of Mars; but by a fantastic coincidence they canceled each other, and he got the right answer. Pasteur demonstrated, by sterilizing organic cultures in sealed flasks, that life does not generate spontaneously from air; but it was lucky that he happened to use an easy-to-kill yeast and not the hay bacillus that another investigator had chosen for the same experiment. We know now that hay bacillus is heat-resistant and grows even after boiling. If Pasteur had used it, his "proof" would have been long a-coming, despite the correctness of his basic idea.

And if the history of science is a very human document, then Gregor Mendel, the father of modern genetics, is a very human scientist. Like Dalton, his conclusions were correct only in general, wrong in detail. Like Arrhenius, he postulated the existence of particles for which there was no experimental evidence—except his; indeed, scientists weren't even prepared for the idea that such particles might exist. Like Garrod, and for the same reasons, he was ignored in his own time. Like Pasteur, he was incredibly lucky in his choice of research material. And like many an investigator before and after his time, his observations reveal the very human tendency to weight the scales in one's own favor, if only subconsciously.

Mendel, an Augustinian monk, had had some training in mathematics and the natural sciences. In addition, he was the son of a farmer, knew the soil, and had a green thumb. Plant hybridization interested him, and he began to read the professional literature. There were many puzzling problems. Crosses between certain species regularly yielded many hybrids with identical traits, for instance; but look what happened when you crossed the hybrids—all kinds of strange new combinations of traits cropped up. The principle of inheritance, if there *was* one, was elusive.

Mendel's basic (and original) idea was that there might be simple mathematical relationships among the characteristic forms of plants in different generations of hybrids. He decided, therefore, to establish some experimental plots in the monastery garden at Brünn, and there raise a number of varieties of peas,[1] hybridize them, count and classify the offspring of each generation, and see whether any mathematical ratios were involved.

Animal and plant breeding had been practiced from the days of the ancient Egyptians, and it was so apparent that children "take after" their parents and that traits "run in families" that the *fact* of inheritance was indisputable. But nobody knew the mechanism. Discoveries identifying the cell as the fundamental unit of life had not been pulled together in any orderly way until the German scientists Schwann and Schleiden did in 1839, and about all that anyone knew for sure in the early years of the nineteenth century was that sperm cells fertilized egg cells. No one knew what went on inside either.

The seventeenth-century idea that sperm contained a "manikin" (a complete but miniature human being) had gone by the boards; and in 1854, when Mendel began his study of inheritance, the prevailing theory was that an "essence" from each vital organ of the parents' bodies somewhat blended to create a new individual. (The belief that a baby gets half its blood from its father and half from its mother is memorialized by such phrases as "blood will tell.") Although it had occurred to various scientists that discrete particles, each affecting different traits, might be passed on from generation to generation, there was no experimental proof. Nor had anyone taken the possibility seriously enough to attempt to prove it.

Mendel had noted that in some varieties of peas, the unripe pods were green while in others they were yellow. Some varieties grew tall, others were dwarf types. Still other pairs of clearly distinct characteristics in different varieties of peas had to do with position of flowers (distributed along the stem or clustered at the top), the form of ripe pods (puffed out or indented), the color of the seed coats (white or gray), the color of the ripe seeds (green or

---

[1] *Not* sweet peas, as is widely believed. The common garden pea, the kind you eat, was his research material.

yellow), and the form of the ripe seeds (smooth or wrinkled). Mendel chose these seven paired characteristics to keep tabs on.

He began with seed that other growers had certified as "pure" (that is, plants grown from it, if self-fertilized, faithfully duplicated the traits of the parental stock)—but just to make sure, he raised plants from it and harvested a crop. Then, by artificial pollination, he crossed varieties in different combinations. The result, for each of the seven paired traits he had chosen to study, was the same: *all* individuals in the first generation took after one parent. It was as if the other parent had had no influence whatever on the result.

Take the cross between smooth and wrinkled peas as an example (figure 1). All the seeds (which are in reality first-generation plants) were smooth. Why? Why weren't there some wrinkled ones, too? Maybe the two traits combined in such a way that one lost its identity, was absorbed by the other; in effect was destroyed. That would certainly be a reasonable conclusion on the basis of this first-generation evidence. But Mendel intended to draw no conclusions until he had followed the various traits in peas through several generations.

He cross-pollinated only to produce the first-generation hybrids. Thereafter—for example, when the first-generation hybrid seeds became plants—they would be allowed to self-pollinate, as is natural for pea plants. In this process, the eggs of a given flower are fertilized by the sperm carried in the pollen of that same flower. And when the second-generation seeds were produced on hybrid plants in this way, Mendel observed that a strange thing had happened. The "lost" traits began to show up! (See figure 2). An

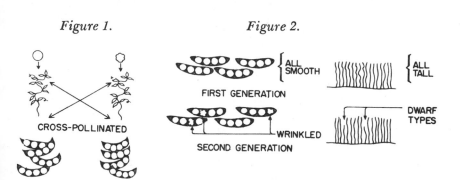

Figure 1.   *Figure 2.*

CROSS-POLLINATED

ALL SMOOTH

FIRST GENERATION

SECOND GENERATION

WRINKLED

ALL TALL

DWARF TYPES

occasional *wrinkled* pea lay alongside the prevailing smooth ones. When the original cross had been between varieties with yellow and green seeds, a few *green* pea seeds were now scattered among the prevailing yellow ones. For characters that show only in mature plants—such as height, position, or color of flowers—the second-generation seeds produced some plants like one of the original "pure" parents, some like the other parent.

Was there a mathematical ratio? Mendel harvested a large number of pods, and in the smooth/wrinkled group he found about 5400 smooth and about 1800 wrinkled seeds. Among those whose original ancestors had been yellow- and green-seeded, he found both types in the second generation in a ratio of about three yellow to one green—approximately 6000 to 2000 in one set of data he recorded. And in the five other paired characteristics he was studying, results were the same: in each case, about 25 per cent of the second-generation plants showed the traits that had appeared to "vanish" in the preceding generation.

Then one trait had *not* absorbed or destroyed the other. One trait must simply have been more "forceful" than the other. The determinants of the two traits had kept their separate identities—except that the weaker one had been submerged, its effect masked. Mendel called the more forceful trait the "dominant" one; the less forceful, the "recessive" one. He expressed the distinction by using capital letters to indicate dominants and lower-case letters to indicate recessives.

Here is the kind of exercise he must have done as he was figuring out the significance of what he had found:

In the cells that give rise to sperm-carrying pollen and also in the cells that produce eggs, let A stand for the determinant for smoothness and a for its alternate, the determinant for wrinkledness. An egg cell has an equal chance of getting A or a, and the same is true for sperm cells. The primary ratio in such cells is therefore $1A$ to $1a$. One or the other will come together in each pairing that brings the next generation into being.

Consider next—as Mendel was having to consider—the possibility that each of the two determinants for a given pair of traits were present as discrete entities (even if invisible) in the cells of the first-generation plants. There would be four ways in which these traits could combine in producing the second-generation off-

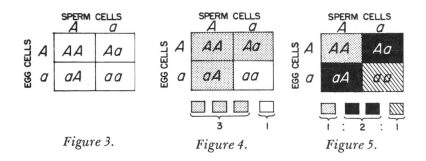

Figure 3.   Figure 4.   Figure 5.

spring. What has since Mendel's day come to be called a "checker-board" diagram graphically indicates (see figure 3) the four com-binations.

Figure 4 repeats the checkerboard, this time with shading over-lying spaces holding dominant A's. All peas of this type would be smooth. Only peas of the aa type would be wrinkled. The second-generation ratio, then, would be $3A$ to $1a$—three-fourths of the progeny smooth, one-fourth wrinkled. This squared very nicely with Mendel's observations upon counting his second-generation crop of pea seeds, 75 per cent of which were smooth and 25 per cent wrinkled.

But 3 : 1 is not an accurate description of their *inherent constitu-tion;* it is descriptive only of their outward appearance. As you see in figure 5, one-fourth of the offspring would be pure for the dominant trait (AA—smoothness), one-fourth would be pure for the recessive trait (aa—wrinkledness), and one-half would be hybrid mixtures of A and a. The correct ratio, then, is $1AA$ to $2Aa$ to $1aa$.

The same principle is illustrated by tossing two coins simultane-ously (see figure 6). There is only one combination that will result in two heads, and only one combination that will result in two tails, but there are two combinations that will result in a heads-and-tails throw (see figure 6). The ratio, here too, is 1 : 2 : 1.

Once Mendel had figured out these mathematical relationships in his first- and second-generation seeds, he knew what he was looking for in subsequent plantings.

He planted the second-generation seeds—the generation, remem-ber, in which 25 per cent of the peas had made a wrinkled come-

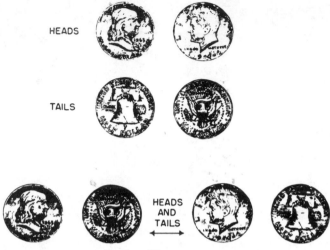

Figure 6.

back; and in the third generation, these wrinkled peas, the "recessives," bred true to type. Furthermore, they did so in all the following years that Mendel planted their descendants.

Of those in the second generation that carried the "dominant" trait—the 75 per cent that looked smooth—only one in three (on the average) bred true to type, and continued thereafter to do so. The diagrams [in figures 7, 8, and 9] show these "pure" lines.

And, on the average, two out of the three that looked smooth did *not* breed true to type, but repeated the pattern of the second generation (see figure 9).

Indeed, inheritance *did* follow an orderly rule. Given a pair of separate and distinct particles to start with, the way in which they would be passed along was statistically predictable.

The logical next question was: Would the same results occur if *two* different traits were crossed? (Let *Aa* stand for smoothness-wrinkledness and *Bb* stand for yellowness-greenness.) The first-generation hybrid plants would then be symbolized as *Aa Bb,* and the eggs of such a plant would be of four kinds: *AB, Ab, aB,* and *ab.* All possible combinations would appear with equal frequency. Perhaps this is a good place to use the examples of coin tossing

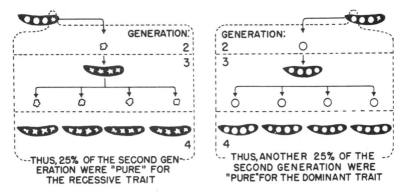

THUS, 25% OF THE SECOND GEN-
ERATION WERE "PURE" FOR
THE RECESSIVE TRAIT

Figure 7.

THUS, ANOTHER 25% OF THE
SECOND GENERATION WERE
"PURE" FOR THE DOMINANT TRAIT

Figure 8.

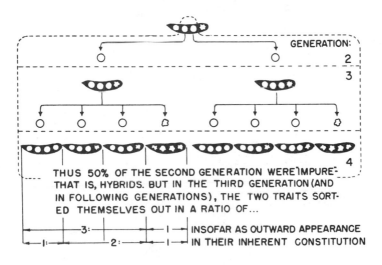

THUS 50% OF THE SECOND GENERATION WERE "IMPURE"-
THAT IS, HYBRIDS. BUT IN THE THIRD GENERATION (AND
IN FOLLOWING GENERATIONS), THE TWO TRAITS SORT-
ED THEMSELVES OUT IN A RATIO OF...

3: 1 — INSOFAR AS OUTWARD APPEARANCE

1: 2: 1 — IN THEIR INHERENT CONSTITUTION

Figure 9.

again, this time a half-dollar and a quarter, flipped at random. As
you see in figure 10, four combinations are possible.

Now, if eggs of the four kinds are fertilized at random by sperms
of the same four kinds, the results can be shown by the checkerboard diagram of sixteen squares in figure 11.

Count the squares that contain either *A*'s or *B*'s: there are nine

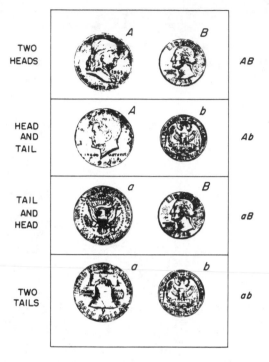

TWO HEADS	*A* *B*	*AB*
HEAD AND TAIL	*A* *b*	*Ab*
TAIL AND HEAD	*a* *B*	*aB*
TWO TAILS	*a* *b*	*ab*

Figure 10.

SPERM CELLS

EGG CELLS	*AB*	*Ab*	*aB*	*ab*
AB	*AABB*	*AABb*	*AaBB*	*AaBb*
Ab	*AABb*	*AAbb*	*AaBb*	*Aabb*
aB	*AaBB*	*AaBb*	*aaBB*	*aaBb*
ab	*AaBb*	*Aabb*	*aaBb*	*aabb*

Figure 11.

of them, and the resultant peas should be smooth and yellow (dominant traits).

Count the squares that contain either *A*'s or *B*'s. There are three of each. Those with *A*'s should express themselves in the crop as smooth green peas; those with *B*'s as wrinkled and yellow.

There remains one square with no capital letters. Peas of this character should be wrinkled and green (recessive traits).

The ratio, as you see, is no longer 1 : 2 : 1 but

9	:	3	:	3	:	1

And how did this paperwork square with the results of a harvested pea crop when Mendel crossed smooth yellow peas with plants yielding wrinkled greens and then raised *their* offspring? Out of 556 ripe peas, he got

Smooth yellow	*Smooth green*	*Wrinkled yellow*	*Wrinkled green*
315	108	101	32

Approximately 9:3:3:1

The inevitable next question, of course, was: Would field tests confirm the statistical prediction of what would happen if one were to cross *three* characters? A first-generation hybrid *Aa Bb Cc* should produce a second-generation ratio of 27 : 9 : 9 : 3 : 9 : 3 : 3 : 1. We'll skip the checkerboard this time, because of its formidable size and complexity[2] but when Mendel crossed hybrid peas that were smooth (*A*), yellow (*B*), and had grayish seed coats (*C*) with plants bearing peas that were wrinkled (*a*), green (*b*), and whose seed coats were white (*c*), his harvest tally of variations added up to a ratio of 27 : 9 : 9 : 3 : 9 : 3 : 3 : 1!

With these and other results that supported his hypothesis, his case was complete. He had grown and harvested peas for twelve years, had kept meticulous records, and now he thought he had

[2] If you want to check on Mendel's math and at the same time show yourself how many different individuals can result from just *three* paired traits, draw a checkerboard with 64 squares and place the following symbols for egg or sperm cells: *ABC, ABc, AbC, Abc, aBC, aBc, abC, abc.* One begins to see, with this exercise, why no human being—whose cells contain not three but thousands of traits—is ever like another human being (unless he has an identical twin).

something worth talking about. So, in 1865, he appeared before the Brünn Society for the Study of Natural Science, read a report of his research, and postulated what have since come to be called the Mendelian laws:

> 1. Inheritance is based on pairs of particulate units, each of which determines specific traits. (He called them "elements"; we call them "genes.") Of each pair, offspring receive *one or the other* from each parent.
> 2. Of each pair of elements acquired by the offspring, one is dominant and the other is recessive. (Mendel called them "antagonistic factors"; we call the two versions of each trait its "alleles.") When both parents contribute a dominant element or when both parents contribute a recessive element (AA or aa), the individual will be "pure" for that trait. If the parent contributes one dominant and one recessive (Aa or aA), the individual will be hybrid, but will look the same as the pure dominant.
> 3. Since these paired elements, whether dominant or recessive, are capable of separating, reappearing in their original form, and pairing differently in later unions, they are obviously not contaminated or altered in any way in the course of their passage from individual to individual.
> 4. If two or more pairs of elements are hybrid in a single plant—Aa and Bb, say—they assort independently in the formation of eggs and sperm; that is, giving AB, Ab, aB, and ab. (This is the law of independent segregation.)

Members of the Society listened politely to Mendel, but insofar as anybody knows asked few questions and engaged in little discussion. It may even be that they sat in embarrassed silence as he proceeded, a suspicion slowly growing that a nice fellow had somehow gotten way off the track.

Mendel's assertion that separate and distinct "elements" of inheritance must exist, despite the fact that he couldn't produce any, was close to asking the Society to accept something on faith. Scientists resist accepting things on faith. There was no evidence for Mendel's hypothesis other than his computations; and his wildly unconventional application of algebra to botany made it difficult for his listeners to understand that those computations *were* the evidence.

Mendel was a careful worker, no doubt about that. And one certainly wouldn't presume to doubt the honesty of a monk. But he'd been raising those peas of his for a long time, and with such single-minded devotion that he might have developed some odd ideas about the implications of his work. Remember, too, that he was little more than a horticultural hobbyist, however dedicated. He lacked a degree, had no University connection, had no previously published research to give him a reputation. Now, if Pasteur had advanced the idea, or Darwin . . .

Anyway, who *really* expects the boy next door to grow up to be President? Surely the gentlemen of the Brünn Society can be forgiven for failing to realize that their modest neighbor had made a brilliant discovery about the fundamental nature of life. They printed his paper in their *Proceedings*—and remembered it as an oddity, if they remembered it at all.

QUESTIONS AND SUGGESTIONS

1. According to the Beadles, why was Mendel's work not accepted? Has this happened to other scientists?

2. The Beadles suggest that had Mendel seen the exceptions to his rules, he might have decided his theory was wrong. When are the exceptions in science important?

3. Outline the organizational methods the Beadles use in writing about the discovery process. Are these methods effective? How does the Beadles' style compare to Watson's in "Finding the Secret of Life"?

4. In analyzing the process of making a discovery, the Beadles imply certain things about the meaning of objectivity in science, the relationship of fact to theory, the role of luck, the relationship between theory, observation, and experiment. Any one of these terms might serve as the subject of another essay.

5. In their introduction to Mendel, the Beadles debunk the view of the scientist as cold and infallible. Compare theirs with James Watson's conception of the scientist. Write an essay persuading your audience that some conception of the scientist is or isn't true.

GERALD HOLTON

Johannes Kepler's Universe:
Its Physics and Metaphysics

For many years, GERALD HOLTON (b. 1922) has taught a famous introductory science course to Harvard undergraduates. His own research has been on the properties of materials under high pressure and in ultrasonics. He has written on the history and philosophy of science and has been active in curricular reform as well. This essay was originally published in the *American Journal of Physics*, and was reprinted in *Thematic Origins of Western Science*.

The important publications of Johannes Kepler (1571–1630) preceded those of Galileo, Descartes, and Newton in time, and in some respects they are even more revealing. And yet, Kepler has been strangely neglected and misunderstood. Very few of his voluminous writings have been translated into English.[1] In this language there has been neither a full biography[2] nor even a major essay on his work in over twenty years. Part of the reason lies in the apparent confusion of incongruous elements—physics and metaphysics, astronomy and astrology, geometry and theology—which characterizes Kepler's work. Even in comparison with Galileo and Newton, Kepler's writings are strikingly different in the *quality* of preoccupation. He is more evidently rooted in a time when animism, alchemy, astrology, numerology, and witchcraft

[1] Books 4 and 5 of the *Epitome of Copernican Astronomy*, and Book 5 of the *Harmonies of the World*, in *Great Books of the Western World* (Chicago: Encyclopaedia Britannica, 1952), Volume 16.

[2] The definitive biography is by the great Kepler scholar Max Caspar, *Johannes Kepler*, Stuttgart: W. Kohlhammer, 1950; the English translation is *Kepler*, trans. and ed. C. Doris Hellman, New York: Abelard-Schuman, 1959. Some useful short essays are in *Johann Kepler, 1571–1630* (A series of papers prepared under the auspices of the History of Science Society in collaboration with the American Association for the Advancement of Science), Baltimore: Williams & Wilkins Co., 1931. [Since this article was written, a number of useful publications on Kepler have appeared—Ed.]

presented problems to be seriously argued. His mode of presentation is equally uninviting to modern readers, so often does he seem to wander from the path leading to the important questions of physical science. Nor is this impression merely the result of the inevitable astigmatism of our historical hindsight. We are trained on the ascetic standards of presentation originating in Euclid, as reestablished, for example, in Books I and II of Newton's *Principia*,[3] and are taught to hide behind a rigorous structure the actual steps of discovery—those guesses, errors, and occasional strokes of good luck without which creative scientific work does not usually occur. But Kepler's embarrassing candor and intense emotional involvement force him to give us a detailed account of his tortuous progress. He still allows himself to be so overwhelmed by the beauty and variety of the world as a whole that he cannot yet persistently limit his attention to the main problems which can in fact be solved. He gives us lengthy accounts of his failures, though sometimes they are tinged with ill-concealed pride in the difficulty of his task. With rich imagination he frequently finds analogies from every phase of life, exalted or commonplace. He is apt to interrupt his scientific thoughts, either with exhortations to the reader to follow a little longer through the almost unreadable account, or with trivial side issues and textual quibbling, or with personal anecdotes or delighted exclamations about some new geometrical relation, a numerological or musical analogy. And sometimes he breaks into poetry or a prayer—indulging, as he puts it, in his "sacred ecstasy." We see him on his pioneering trek, probing for the firm ground on which our science could later build, and often led into regions which we now know to be unsuitable marshland.

These characteristics of Kepler's style are not merely idiosyncrasies. They mirror the many-sided struggle attending the rise of modern science in the early seventeenth century. Conceptions which we might now regard as mutually exclusive are found to operate side-by-side in his intellectual make-up. A primary aim of this essay is to identify those disparate elements and to show that

3 But Newton's *Opticks*, particularly in the later portions, is rather reminiscent of Kepler's style. In Book II, Part IV, Observation 5, there is, for example, an attempt to associate the parts of the light spectrum with the "differences of the lengths of a monochord which sounds the tones in an eight."

in fact much of Kepler's strength stems from their juxtaposition. We shall see that when his physics fails, his metaphysics comes to the rescue; when a mechanical model breaks down as a tool of explanation, a mathematical model takes over; and at its boundary in turn there stands a theological axiom. Kepler set out to unify the classical picture of the world, one which was split into celestial and terrestrial regions, through the concept of a universal physical *force;* but when this problem did not yield to physical analysis, he readily returned to the devices of a unifying *image,* namely, the central sun ruling the world, and of a unifying *principle,* that of all-pervading mathematical harmonies. In the end he failed in his initial project of providing the mechanical explanation for the observed motions of the planets, but he succeeded at least in throwing a bridge from the old view of the world as unchangeable *cosmos* to the new view of the world as the playground of dynamic and mathematical laws. And in the process he turned up, as if it were by accident, those clues which Newton needed for the eventual establishment of the new view.

Toward a Celestial Machine

A sound instinct for physics and a commitment to neo-Platonic metaphysics—these are Kepler's two main guides which are now to be examined separately and at their point of merger. As to the first, Kepler's genius in physics has often been overlooked by critics who were taken aback by his frequent excursions beyond the bounds of science as they came to be understood later, although his *Dioptrice* (1611) and his mathematical work on infinitesimals (in *Nova Stereometria,* 1615) and on logarithms (*Chilias Logarithmorum,* 1624) have direct appeal for the modern mind. But even Kepler's casually delivered opinions often prove his insight beyond the general state of knowledge of his day. One example is his creditable treatment of the motion of projectiles on the rotating earth, equivalent to the formulation of the superposition principle of velocities.[4] Another is his opinion of the *perpetuum mobile:*

> As to this matter, I believe one can prove with very good reasons that neither any never-ending motion nor the quad-

[4] Letter to David Fabricius, October 11, 1605.

rature of the circle—two problems which have tortured great minds for ages—will ever be encountered or offered by nature.[5]

But, of course, on a large scale, Kepler's genius lies in his early search for a physics of the solar system. He is the first to look for *a universal physical law based on terrestrial mechanics* to comprehend the whole universe in its quantitative details. In the Aristotelian and Ptolemaic world schemes, and indeed in Copernicus's own, the planets moved in their respective orbits by laws which were either purely mathematical or mechanical in a nonterrestrial sense. As Goldbeck reminds us, Copernicus himself still warned to keep a clear distinction between celestial and merely terrestrial phenomena, so as not to "attribute to the celestial bodies what belongs to the earth."[6] This crucial distinction disappears in Kepler from the beginning. In his youthful work of 1596, the *Mysterium Cosmographicum*, a single geometrical device is used to show the necessity of the observed orbital arrangement of all planets. In this respect, the earth is treated as being an equal of the other planets.[7] In the words of Otto Bryk,

5 Letter to Herwart von Hohenburg, March 26, 1598, i.e., seven years before Stevinus implied the absurdity of perpetual motion in the *Hypomnemata Mathematica* (Leyden, 1605). Some of Kepler's most important letters are collected in Max Caspar and Walther von Dyck, *Johannes Kepler in seinen Briefen*, Munich and Berlin: R. Oldenbourg, 1930. A more complete collection in the original languages is to be found in Vols. 13–15 of the modern edition of Kepler's collected works, *Johannes Keplers gesammelte Werke*, ed. von Dyck and Caspar, Munich: C. H. Beck, 1937 and later. In the past, these letters appear to have received insufficient attention in the study of Kepler's work and position. (The present English translations of all quotations from them are the writer's.) Excerpts from some letters were also translated in Carola Baumgardt, *Johannes Kepler*, New York: Philosophical Library, 1951.

6 Ernst Goldbeck, *Abhandlungen zur Philosophie und ihrer Geschichte, Keplers Lehre von der Gravitation* (Halle: Max Niemeyer, 1896), Volume VI—a useful monograph demonstrating Kepler's role as a herald of mechanical astronomy. The reference is to *De Revolutionibus*, first edition, p. 3. [The main point, which it would be foolhardy to challenge, is that in the description of phenomena Copernicus still on occasion treated the earth differently from other planets.]

7 In Kepler's Preface to his *Dioptrice* (1611) he calls his early *Mysterium Cosmographicum* "a sort of combination of astronomy and Euclid's Geometry," and describes the main features as follows: "I took the dimensions of the planetary orbits according to the astronomy of Copernicus, who makes the sun im-

The central and permanent contribution lies in this, that for the first time the whole world structure was subjected to a single law of construction—though not a force law such as revealed by Newton, and only a non-causative relationship between spaces, but nevertheless one single law.[8]

Four years later Kepler meets Tycho Brahe and from him learns to respect the power of precise observation. The merely approximate agreement between the observed astronomical facts and the scheme laid out in the *Mysterium Cosmographicum* is no longer satisfying. To be sure, Kepler always remained fond of his work, and in the *Dissertatio cum Nuncio Sidereo* (1610) even hoped that Galileo's newly-found moons of Jupiter would help to fill in one of the gaps left in his geometrical model. But with another part of his being Kepler knows that an entirely different approach is wanted. And here Kepler turns to the new conception of the universe. While working on the *Astronomia Nova* in 1605, Kepler lays out his program:

I am much occupied with the investigation of the physical causes. My aim in this is to show that the celestial machine is to be likened not to a divine organism but rather to a clockwork . . . , insofar as nearly all the manifold movements are carried out by means of a single, quite simple magnetic force, as in the case of a clockwork all motions [are caused] by a simple weight. Moreover I show how this physical conception is to be presented through calculation and geometry.[9]

mobile in the center, and the earth movable both round the sun and upon its own axis; and I showed that the differences of their orbits corresponded to the five regular Pythagorean figures, which had been already distributed by their author among the elements of the world, though the attempt was admirable rather than happy or legitimate. . . ." The scheme of the five circumscribed regular bodies originally represented to Kepler the *cause* of the observed number (and orbits) of the planets: "*Habes rationem numeri planetarium.*"

[8] Johannes Kepler, *Die Zusammenklänge der Welten*, Otto J. Bryk, trans. and ed. (Jena: Diederichs, 1918), p. xxiii.

[9] Letter to Herwart von Hohenburg, February 10, 1605. At about the same time he writes in a similar vein to Christian Severin Longomontanus concerning the relation of astronomy and physics: "I believe that both sciences are so closely interlinked that the one cannot attain completion without the other."

The celestial machine, driven by a single terrestrial force, in the image of a clockwork! This is indeed a prophetic goal. Published in 1609, the *Astronomia Nova* significantly bears the subtitle *Physica Coelestis*. The book is best known for containing Kepler's First and Second Laws of planetary motion, but it represents primarily a search for one universal force law to explain the motions of planets—Mars in particular—as well as gravity and the tides. This breathtaking conception of unity is perhaps even more striking than Newton's, for the simple reason that Kepler had no predecessor.

The Physics of the Celestial Machine

Kepler's first recognition is that forces between bodies are caused not by their relative positions or their geometrical arrangements, as was accepted by Aristotle, Ptolemy, and Copernicus, but by mechanical interactions between the material objects. Already in the *Mysterium Cosmographicum* (Chapter 17) he announced "*Nullum punctum, nullum centrum grave est*," and he gave the example of the attraction between a magnet and a piece of iron. In William Gilbert's *De Magnete* (1600), published four years later, Kepler finds a careful explanation that the action of magnets seems to come from pole points, but must be attributed to the parts of the body, not the points.

In the spirited *Objections* which Kepler appended to his own translation of Aristotle's Περὶ οὐρανοῦ, he states epigrammatically "*Das Mittele is nur ein Düpfflin*," and he elaborates as follows:

> How can the earth, or its nature, notice, recognize and seek after the center of the world which is only a little point [*Düpfflin*]—and then go toward it? The earth is not a hawk, and the center of the world not a little bird; it [the center] is also not a magnet which could attract the earth, for it has no substance and therefore cannot exert a force.

In the Introduction to the *Astronomia Nova*, which we shall now consider in some detail, Kepler is quite explicit:

> A mathematical point, whether it be the center of the world or not, cannot move and attract a heavy object. . . . Let the [Aristotelian] physicists prove that such a force is to

be associated with a point, one which is neither corporeal nor recognisable as anything but a pure reference [mark].

Thus what is needed is a "true doctrine concerning gravity"; the axioms leading to it include the following:

> Gravitation consists in the mutual bodily striving among related bodies toward union or connection; (of this order is also the magnetic force).

This premonition of universal gravitation is by no means an isolated example of lucky intuition. Kepler's feeling for the physical situation is admirably sound, as shown in additional axioms:

> If the earth were not round, a heavy body would be driven not everywhere straight toward the middle of the earth, but toward different points from different places.
>
> If one were to transport two stones to any arbitrary place in the world, closely together but outside the field of force [extra orbe virtutis] of a third related body, then those stones would come together at some intermediate place similar to two magnetic bodies, the first approaching the second through a distance which is proportional to the mass [moles] of the second.

And after this precursor of the principle of conservation of momentum, there follows the first attempt at a good explanation for the tides in terms of a force of attraction exerted by the moon.

But the Achilles' heel of Kepler's celestial physics is found in the very first "axiom," in his Aristotelian conception of the law of inertia, where inertia is identified with a tendency to come to rest —causa privativa motus:

> Outside the field of force of another related body, every bodily substance, insofar as it is corporeal, by nature tends to remain at the same place at which it finds itself.[10]

This axiom deprives him of the concepts of mass and force in useful form—the crucial tools needed for shaping the celestial

---

[10] Previously, Kepler discussed the attraction of the moon in a letter to Herwart, January 2, 1607. The relative motion of two isolated objects and the concept of inertia are treated in a letter to D. Fabricius, October 11, 1605. On the last subject see Alexandre Koyré, "Galileo and the Scientific Revolution of the Seventeenth Century," The Philosophical Review, 52, No. 4: 344–345, 1943.

metaphysics of the ancients into the celestial physics of the moderns. Without these concepts, Kepler's world machine is doomed. He has to provide separate forces for the propulsion of planets tangentially along their paths and for the radial component of motion.

Moreover, he assumed that the force which reaches out from the sun to keep the planets in tangential motion falls inversely with the increasing distance. The origin and the consequences of this assumption are very interesting. In Chapter 20 of the *Mysterium Cosmographicum,* he speculated casually why the sidereal periods of revolution on the Copernican hypothesis should be larger for the more distant planets, and what force law might account for this:

> We must make one of two assumptions: either the forces of motion [*animae motrices*] [are inherent in the planets] and are feebler the more remote they are from the sun, or there is only one *anima motrix* at the center of the orbits, that is, in the sun. It drives the more vehemently the closer the [moved] body lies; its effect on the more distant bodies is reduced because of the distance [and the corresponding] decrease of the impulse. Just as the sun contains the source of light and the center of the orbits, even so can one trace back to this same sun life, motion and the soul of the world. . . . Now let us note how this decrease occurs. To this end we will assume, as is very probable, that the moving effect is weakened through spreading from the sun in the same manner as light.

This suggestive image—with its important overtones which we shall discuss below—does, however, not lead Kepler to the inverse-square law of force, for he is thinking of the spreading of light *in a plane,* corresponding to the plane of planetary orbits. The decrease of light intensity is therefore associated with the linear increase in circumference for more distant orbits! In his pre-Newtonian physics, where force is proportional not to acceleration but to velocity, Kepler finds a ready use for the inverse first-power law of gravitation. It is exactly what he needs to explain his observation that the speed of a planet in its elliptical orbit decreases linearly with the increase of the planet's distance from the sun. Thus Kepler's Second Law of Planetary Motion—which he actually discovered *before* the so-called First and Third laws—finds a partial physical explanation in joining several erroneous postulates.

In fact, it is clear from the context that these postulates originally suggested the Second Law to Kepler.[11] But not always is the final outcome so happy. Indeed, the hypothesis concerning the physical forces acting on the planet seriously delays Kepler's progress toward the law of elliptical orbits (First Law). Having shown that "the path of the planet [Mars] is not a circle but an oval figure," he attempts (Chapter 45, *Astronomia Nova*) to find the details of a physical force law which would explain the "oval" path in a quantitative manner. But after ten chapters of tedious work he has to confess that "the physical causes in the forty-fifth chapter thus go up in smoke." Then in the remarkable fifty-seventh chapter, a final and rather desperate attempt is made to formulate a force law. Kepler even dares to entertain the notion of combined magnetic influences and animal forces [*vis animalia*] in the planetary system. Of course, the attempt fails. The accurate clockwork-like celestial machine cannot be constructed.

To be sure, Kepler does not give up his conviction that a universal force exists in the universe, akin to magnetism. For example, in Book 4 of the *Epitome of Copernican Astronomy* (1620), we encounter the picture of a sun as a spherical magnet with one pole at the center and the other distributed over its surface. Thus a planet, itself magnetized like a bar magnet with a fixed axis, is alternately attracted to and repelled from the sun in its elliptical orbit. This is to explain the radial component of planetary motion. The tangential motion has been previously explained (in Chapter 34, *Astronomia Nova*) as resulting from the drag or torque which magnetic lines of force from the rotating sun are supposed to exert on the planet as they sweep over it. But the picture remains qualitative and incomplete, and Kepler does not return to his original plan to "show how this physical conception is to be presented through calculation and geometry." [See foot-

11 Not only the postulates but also some of the details of their use in the argument were erroneous. For a short discussion of this concrete illustration of Kepler's use of physics in astronomy, see John L. E. Dreyer, *History of the Planetary System from Thales to Kepler* (New York: Dover Publications, 1953), second edition, pp. 387–399. A longer discussion is in Max Caspar, *Johannes Kepler, neue Astronomie* (Munich and Berlin: R. Oldenbourg, 1929), pp. 3–66.

note 9.] Nor does his long labor bring him even a fair amount of recognition. Galileo introduces Kepler's work into his discussion on the world systems only to scoff at Kepler's notion that the moon affects the tides,[12] even though Tycho Brahe's data and Kepler's work based on them had shown that the Copernican scheme which Galileo was so ardently upholding did not correspond to the experimental facts of planetary motion. And Newton manages to remain strangely silent about Kepler throughout Books I and II of the *Principia*, by introducing the Third Law anonymously as "the phenomenon of the 3/2th power" and the First and Second Laws as "the *Copernican* hypothesis."[13] Kepler's three laws have come to be treated as essentially empirical rules. How far removed this achievement was from his original ambition!

Kepler's First Criterion of Reality: The Physical Operations of Nature

Let us now set aside for a moment the fact that Kepler failed to build a mechanical model of the universe, and ask why he undertook the task at all. The answer is that Kepler (rather like Galileo) was trying to establish a new philosophical interpretation for "reality." Moreover, he was quite aware of the novelty and difficulty of the task.

In his own words, Kepler wanted to "provide a philosophy or physics of celestial phenomena in place of the theology or metaphysics of Aristotle."[14] Kepler's contemporaries generally regarded

[12] Giorgio de Santillana, ed., *Dialogue on the Great World Systems* (Chicago: University of Chicago Press, 1953), p. 469. However, an oblique compliment to Kepler's Third Law may be intended in a passage on p. 286.

[13] Florian Cajori, ed., *Newton's Principia: Motte's Translation Revised* (Berkeley: University of California Press, 1946), pp. 394–395. In Book III, Newton remarks concerning the fact that the Third Law applies to the moons of Jupiter: "This we know from astronomical observations." At last, on page 404, Kepler is credited with having "first observed" that the 3/2th power law applies to the "five primary planets" and the earth. Newton's real debt to Kepler was best summarized in his own letter to Halley, July 14, 1686: "But for the duplicate proportion [the inverse-square law of gravitation] I can affirm that I gathered it from Kepler's theorem about twenty years ago."

[14] Letter to Johann Brengger, October 4, 1607. This picture of a man struggling to emerge from the largely Aristotelian tradition is perhaps as significant as

his intention of putting laws of physics into astronomy as a new and probably pointless idea. Even Michael Mästlin, Kepler's own beloved teacher, who had introduced Kepler to the Copernican theory, wrote him on October 1, 1616:

> Concerning the motion of the moon you write you have traced all the inequalities to physical causes; I do not quite understand this. I think rather that here one should leave physical causes out of account, and should explain astronomical matters only according to astronomical method with the aid of astronomical, not physical, causes and hypotheses. That is, the calculation demands astronomical bases in the field of geometry and arithmetic. . . .

The difference between Kepler's conception of the "physical" problems of astronomy and the methodology of his contemporaries reveals itself clearly in the juxtaposition of representative letters by the two greatest astronomers of the time—Tycho Brahe and Kepler himself. Tycho, writing to Kepler on December 9, 1599, repeats the preoccupation of two millennia of astronomical speculations:

> I do not deny that the celestial motions achieve a certain symmetry [through the Copernican hypothesis], and that there are reasons why the planets carry through their revolutions around this or that center at different distances from the earth or the sun. However, the harmony or regularity of the scheme is to be discovered only a posteriori. . . . And even if it should appear to some puzzled and rash fellow that the superposed circular movements on the heavens yield sometimes angular or other figures, mostly elongated ones, then it happens accidentally, and reason recoils in horror from this assumption. For one must compose the revolutions of celestial objects definitely from circular motions; otherwise they could not come back on the same path eter-

---

the usual one of Kepler as Copernican in a Ptolemaic world. Nor was Kepler's opposition, strictly speaking, Ptolemaic any longer. For this we have Kepler's own opinion (*Harmonice Mundi*, Book 3): "First of all, readers should take it for granted that among astronomers it is nowadays agreed that all planets circulate around the sun . . . ," meaning of course the system not of Copernicus but of Tycho Brahe, in which the earth was fixed and the moving sun served as center of motion for the other planets.

nally in equal manner, and an eternal duration would be impossible, not to mention that the orbits would be less simple, and irregular, and unsuitable for scientific treatment.

This manifesto of ancient astronomy might indeed have been subscribed to by Pythagoras, Plato, Aristotle, and Copernicus himself. Against it, Kepler maintains a new stand. Writing to D. Fabricius on August 1, 1607, he sounds the great new *leitmotif* of astronomy: *"The difference consists only in this, that you use circles, I use bodily forces."* And in the same letter, he defends his use of the ellipse in place of the superposition of circles to represent the orbit of Mars:

> When you say it is not to be doubted that all motions occur on a perfect circle, then this is false for the composite, i.e., the real motions. According to Copernicus, as explained, they occur on an orbit distended at the sides, whereas according to Ptolemy and Brahe on spirals. But if you speak of components of motion, then you speak of something existing in thought; i.e., something that is not there in reality. For nothing courses on the heavens except the planetary bodies themselves—no orbs, no epicycles. . . .

This straightforward and modern-sounding statement implies that behind the word "real" stands "mechanical," that for Kepler the real world is the world of objects and of their mechanical interactions in the sense which Newton used; e.g., in the preface to the *Principia:*

> Then from these [gravitational] forces, by other propositions which are also mathematical, I deduce the motions of the planets, the comets, the moon, and the sea. I wish we could derive the rest of the phenomena of nature by the same kind of reasoning from mechanical principles. . . .[15]

Thus we are tempted to see Kepler as a natural philosopher of the mechanistic-type later identified with the Newtonian disciples. But this is deceptive. Particularly after the failure of the program of the *Astronomia Nova,* another aspect of Kepler asserted itself. Though he does not appear to have been conscious of it, he never

[15] Cajori, *op. cit.,* p. xviii.

resolved finally whether the criteria of reality are to be sought on the *physical* or the *metaphysical* level. The words "real" or "physical" themselves, as used by Kepler, carry two interpenetrating complexes of meaning. Thus on receiving Mästlin's letter of October 1, 1616, Kepler jots down in the margin his own definition of "physical":

> I call my hypotheses for two reasons. . . . My aim is to assume only those things of which I do not doubt they are real and consequently physical, where one must refer to the nature of the heavens, not the elements. When I dismiss the perfect eccentric and the epicycle, I do so because they are purely geometrical assumptions, for which a corresponding body in the heavens does not exist. The second reason for my calling my hypotheses physical is this . . . I prove that the irregularity of the motion [of planets] corresponds to the nature of the planetary sphere; i.e., is physical.

This throws the burden on the *nature* of heavens, the *nature* of bodies. How, then, is one to recognize whether a postulate or conception is in accord with the nature of things?

This is the main question, and to it Kepler has at the same time two very different answers, emerging, as it were, from the two parts of his soul. We may phrase one of the two answers as follows: *the physically real world, which defines the nature of things, is the world of phenomena explainable by mechanical principles.* This can be called Kepler's first criterion of reality, and assumes the possibility of formulating a sweeping and consistent dynamics which Kepler only sensed but which was not to be given until Newton's *Principia.* Kepler's other answer, to which he keeps returning again and again as he finds himself rebuffed by the deficiencies of his dynamics, and which we shall now examine in detail, is this: *the physically real world is the world of mathematically expressed harmonies which man can discover in the chaos of events.*

Kepler's Second Criterion of Reality: The Mathematical Harmonies of Nature

Kepler's failure to construct a *Physica Coelestis* did not damage his conception of the astronomical world. This would be strange indeed in a man of his stamp if he did not have a ready alternative to the mechanistic point of view. Only rarely does he seem to have been really uncomfortable about the poor success of the latter, as when he is forced to speculate how a soul or an inherent intelligence would help to keep a planet on its path. Or again, when the period of rotation of the sun which Kepler had postulated in his physical model proved to be very different from the actual rotation as first observed through the motion of sunspots, Kepler was characteristically not unduly disturbed. The truth is that despite his protestations, Kepler was not as committed to mechanical explanations of celestial phenomena as was, say, Newton. He had another route open to him.

His other criterion, his second answer to the problem of physical reality, stemmed from the same source as his original interest in astronomy and his fascination with a universe describable in mathematical terms, namely from a frequently acknowledged metaphysics rooted in Plato and neo-Platonists such as Procius Diadochus. It is the criterion of *harmonious regularity in the descriptive laws of science.* One must be careful not to dismiss it either as just a reappearance of an old doctrine or as an aesthetic requirement which is still recognized in modern scientific work; Kepler's conception of what is "harmonious" was far more sweeping and important than either.

A concrete example is again afforded by the Second Law, the "Law of Equal Areas." To Tycho, Copernicus, and the great Greek astronomers, the harmonious regularity of planetary behavior was to be found in the uniform motion in component circles. But Kepler recognized the orbits—after a long struggle—as ellipsi on which planets move in a nonuniform manner. The figure is lopsided. The speed varies from point to point. And yet, nestled within this double complexity is hidden a harmonious regularity which transports its ecstatic discoverer—namely, the fact that a constant area is swept out in equal intervals by a line from the focus of the ellipse, where the sun is, to the planet on the ellipse. For Kepler, the law is harmonious in three separate senses.

First, *it is in accord with experience.* Whereas Kepler, despite long and hard labors, had been unable to fit Tycho's accurate observations on the motion of Mars into a classical scheme of superposed circles, the postulate of an elliptical path fitted the observations at once. Kepler's dictum was: "harmonies must accommodate experience."[16] How difficult it must have been for Kepler, a Pythagorean to the marrow of his bones, to forsake circles for ellipsi! For a mature scientist to find in his own work the need for abandoning his cherished and ingrained preconceptions, the very basis of his previous scientific work, in order to fulfill the dictates of quantitative experience—this was perhaps one of the great sacrificial acts of modern science, equivalent in recent scientific history to the agony of Max Planck. Kepler clearly drew the strength for this act from the belief that it would help him to gain an even deeper insight into the harmony of the world.

The second reason for regarding the law as harmonious is its reference to, or discovery of, a *constancy,* although no longer a constancy simply of angular velocity but of areal velocity. The typical law of ancient physical science had been Archimedes' law of the lever: a relation of direct observables in static configuration. Even the world systems of Copernicus and of Kepler's *Mysterium Cosmographicum* still had lent themselves to visualization in terms of a set of fixed concentric spheres. And we recall that Galileo never made use of Kepler's ellipsi, but remained to the end a true follower of Copernicus who had said "the mind shudders" at the supposition of noncircular nonuniform celestial motion, and "it would be unworthy to suppose such a thing in a Creation constituted in the best possible way."

With Kepler's First Law and the postulation of elliptical orbits, the old simplicity was destroyed. The Second and Third Laws established the physical law of constancy as an ordering principle in a changing situation. Like the concepts of momentum and caloric in later laws of constancy, areal velocity itself is a concept far removed from the immediate observables. It was therefore a bold step to search for harmonies beyond both perception and preconception.

Thirdly, the law is harmonious also in a grandiose sense: the fixed point of reference in the Law of Equal Areas, the "center" of planetary motion, is the center of the *sun itself,* whereas even in the Copernican scheme the sun was a little off the center of

16 Quoted in Kepler, *Weltharmonik,* ed. Max Caspar (Munich and Berlin: R. Oldenbourg, 1939), p. 55.

planetary orbits. With this discovery Kepler makes the planetary system at last truly heliocentric, and thereby satisfies his instinctive and sound demand for some material object as the "center" to which ultimately the physical effects that keep the system in orderly motion must be traced.

A Heliocentric and Theocentric Universe

For Kepler, the last of these three points is particularly exciting. The sun at its fixed and commanding position at the center of the planetary system matches the picture which always rises behind Kepler's tables of tedious data—the picture of a centripetal universe, directed toward and guided by the *sun* in its manifold roles: as the *mathematical* center in the description of celestial motions; as the central *physical* agency for assuring continued motion; and above all as the *metaphysical* center, the temple of the Deity. The three roles are in fact inseparable. For granting the special simplicity achieved in the description of planetary motions in the heliocentric system, as even Tycho was willing to grant, and assuming also that each planet must experience a force to drag it along its own constant and eternal orbit, as Kepler no less than the Scholastics thought to be the case, then it follows that the common need is supplied from what is common to all orbits; i.e., their common center, and this source of eternal constancy itself must be constant and eternal. Those, however, are precisely the unique attributes of the Deity.

Using his characteristic method of reasoning on the basis of archetypes, Kepler piles further consequences and analogies on this argument. The most famous is the comparison of the world-sphere with the Trinity: the sun, being at the center of the sphere and thereby antecedent to its two other attributes, surface and volume, is compared to God the Father. With variations the analogy occurs many times throughout Kepler's writings, including many of his letters. The image haunts him from the very beginning (e.g., Chapter 2, *Mysterium Cosmographicum*) and to the very end. Clearly, it is not sufficient to dismiss it with the usual phrase "sunworship."[17] At the very least, one would have to allow

17 E.g., Edwin Arthur Burtt, *The Metaphysical Foundations of Modern Science* (London: Routledge & Kegan Paul, 1924 and 1932), p. 47 ff.

that the exuberant Kepler is a worshipper of the whole solar system in all its parts.

The power of the sun-image can be traced to the acknowledged influence on Kepler by neo-Platonists such as Proclus (fifth century) and Witelo (thirteenth century). At the time it was current neo-Platonic doctrine to identify light with "the source of all existence" and to hold that "space and light are one."[18] Indeed, one of the main preoccupations of the sixteenth-century neo-Platonists had been, to use a modern term, the transformation properties of space, light, and soul. Kepler's discovery of a truly heliocentric system is not only in perfect accord with the conception of the sun as a ruling entity, but allows him, for the first time, to focus attention on the sun's position through argument from physics.

In the medieval period the "place" for God, both in Aristotelian and in neo-Platonic astronomical metaphysics, had commonly been either beyond the last celestial sphere or else all of space; for only those alternatives provided for the Deity a "place" from which all celestial motions were equivalent. But Kepler can adopt a third possibility: in a truly heliocentric system God can be brought back into the solar system itself, so to speak, enthroned at the fixed and common reference object which coincides with the source of light and with the origin of the physical forces holding the system together. In the *De Revolutionibus* Copernicus had glimpsed part of this image when he wrote, after describing the planetary arrangement:

> In the midst of all, the sun reposes, unmoving. Who, indeed, in this most beautiful temple would place the light-giver in any other part than that whence it can illumine all other parts.

But Copernicus and Kepler were quite aware that the Copernican sun was not quite "in the midst of all"; hence Kepler's delight when, as one of his earliest discoveries, he found that orbital planes of all planets intersect at the sun.

[18] For a recent analysis of neo-Platonic doctrine, which regrettably omits a detailed study of Kepler, see Max Jammer, *Concepts of Space* (Cambridge: Harvard University Press, 1954), p. 37 ff. Neo-Platonism in relation to Kepler is discussed by Thomas S. Kuhn, *The Copernican Revolution*, Cambridge: Harvard University Press, 1957.

The threefold implication of the heliocentric image as mathematical, physical, and metaphysical center helps to explain the spell it casts on Kepler. As Wolfgang Pauli has pointed out in a highly interesting discussion of Kepler's work as a case study in "the origin and development of scientific concepts and theories," here lies the motivating clue: "It is because he sees the sun and planets against the background of this fundamental image [archetypische Bild] that he believes in the heliocentric system with religious fervor"; it is this belief "which causes him to search for the true laws concerning the proportion in planetary motion. . . ."[19]

To make the point succinctly, we may say that in its final version *Kepler's physics of the heavens is heliocentric in its kinematics, but theocentric in its dynamics,* where harmonies based in part on the properties of the Deity serve to supplement physical laws based on the concept of specific quantitative forces. This brand of physics is most prominent in Kepler's last great work, the *Harmonice Mundi* (1619). There the so-called Third Law of planetary motion is announced without any attempt to deduce it from mechanical principles, whereas in the *Astronomia Nova* magnetic forces had driven—no, obsessed—the planets. As in his earliest work, he shows that the phenomena of nature exhibit underlying mathematical harmonies. Having not quite found the mechanical gears of the world machine, he can at least give its equations of motion.

The Source of Kepler's Harmonies

Unable to identify Kepler's work in astronomy with physical science in the modern sense, many have been tempted to place him on the other side of the imaginary dividing line between classical and modern science. Is it, after all, such a large step from the harmonies which the ancients found in circular motion and rational numbers to the harmonies which Kepler found in elliptical mo-

[19] Wolfgang Pauli, *Der Einfluss archetypischer Vorstellungen auf die Bildung naturwissenschaftlicher Theorien bei Kepler,* in *Naturerklärung und Psyche* (Zurich: Rascher Verlag, 1952), p. 129.

An English translation of Jung and Pauli is *The Interpretation of Nature and the Psyche,* trans. R. F. C. Hull and Priscilla Silz, New York: Pantheon Books, 1955.

tions and exponential proportions? Is it not merely a generalization of an established point of view? Both answers are in the negative. For the ancients and for most of Kepler's contemporaries, the
hand of the Deity was revealed in nature through laws which, if
not qualitative, were harmonious in an essentially self-evident
way; the axiomatic simplicity of circles and spheres and integers
itself proved their deistic connection. But Kepler's harmonies reside in the very fact that the relations *are quantitative,* not in some
specific simple *form* of the quantitative relations.

*It is exactly this shift which we can now recognize as one point
of breakthrough toward the later, modern conception of mathematical law in science.* Where in classical thought the quantitative
actions of nature were limited by a few necessities, the new attitude, whatever its metaphysical motivation, opens the imagination
to an infinity of possibilities. As a direct consequence, where in
classical thought the quantitative results of experience were used
largely to fill out a specific pattern by a priori necessity, the new
attitude permits the results of experience to reveal in themselves
whatever pattern nature has in fact chosen from the infinite set of
possibilities. Thus the seed is planted for the general view of most
modern scientists, who find the world harmonious in a vague aesthetic sense because the mind can find, inherent in the chaos of
events, order framed in mathematical laws—of whatever form they
may be. As has been aptly said about Kepler's work:

> Harmony resides no longer in numbers which can be
> gained from arithmetic without observation. Harmony is
> also no longer the property of the circle in higher measure
> than the ellipse. Harmony is present when a multitude of
> phenomena is regulated by the unity of a mathematical law
> which expresses a cosmic idea.[20]

Perhaps it was inevitable in the progress of modern science that
the harmony of mathematical law should now be sought in aesthetics rather than in metaphysics. But Kepler himself would
have been the last to propose or accept such a generalization.
The ground on which he postulated that harmonies reside in the
quantitative properties of nature lies in the same metaphysics
which helped him over the failure of his physical dynamics of the

[20] Hedwig Zaiser, *Kepler als Philosoph* (Stuttgart: E. Suhrkamp, 1932), p. 47.

solar system. Indeed, the source is as old as natural philosophy itself: *the association of quantity per se with Deity*. Moreover, as we can now show, Kepler held that man's ability to discover harmonies, and therefore reality, in the chaos of events is due to a direct connection between ultimate reality; namely, God, and the mind of man.

In an early letter, Kepler opens to our view this mainspring of his life's work:

> May God make it come to pass that my delightful speculation [the *Mysterium Cosmographicum*] have everywhere among reasonable men fully the effect which I strove to obtain in the publication; namely, that the belief in the creation of the world be fortified through this external support, that thought of the creator be recognized in its nature, and that His inexhaustible wisdom shine forth daily more brightly. Then man will at last measure the power of his mind on the true scale, and will realize that *God, who founded everything in the world according to the norm of quantity, also has endowed man with a mind which can comprehend these norms*. For as the eye for color, the ear for musical sounds, so is the mind of man created for the perception not of any arbitrary entities, but rather of quantities; the mind comprehends a thing the more correctly the closer the thing approaches toward pure quantity as its origin.[21]

On a superficial level, one may read this as another repetition of the old Platonic principle ὁ νεὸς ἀεὶ γεωμετεῖ; and of course Kepler does believe in "the creator, the true first cause of geometry, who, as Plato says, always geometrizes."[22] Kepler is indeed a Platonist, and even one who is related at the same time to both neo-Platonic traditions—which one might perhaps better identify as

[21] Letter to Mästlin, April 19, 1597. (Italics supplied.) The "numerological" component of modern physical theory is in fact a respectable offspring from this respectable antecedent. For example, see Niels Bohr, *Atomic Theory and the Description of Nature* (New York: Macmillan Co., 1934), pp. 103–104: "This interpretation of the atomic number [as the number of orbital electrons] may be said to signify an important step toward the solution of one of the boldest dreams of natural science, namely, to build up an understanding of the regularities of nature upon the consideration of pure number."

[22] *Harmonice Mundi*, Book 3.

the neo-Platonic and the neo-Pythagorean—that of the mathematical physicists like Galileo and that of the mathematical mysticism of the Florentine Academy. But Kepler's God has done more than build the world on a mathematical model; he also specifically created man with a mind which "carries in it concepts built on the category of quantity," *in order that man may directly communicate with the Deity:*

> Those laws [which govern the material world] lie within the power of understanding of the human mind; God wanted us to perceive them when he created us in His image in order that we may take part in His own thoughts. . . . Our knowledge [of numbers and quantities] is of the same kind as God's, at least insofar as we can understand something of it in this mortal life.[23]

The procedure by which one apprehends harmonies is described quite explicitly in Book 4, Chapter 1, of *Harmonice Mundi*. There are two kinds of harmonies; namely, those in sense phenomena, as in music, and in "pure" harmonies such as are "constructed of mathematical concepts." The feeling of harmony arises when there occurs a matching of the perceived order with the corresponding innate archetype [*archetypus, Urbild*]. The archetype itself is part of the mind of God and was impressed on the human soul by the Deity when He created man in His image. The kinship with Plato's doctrine of ideal forms is clear. But whereas the latter, in the usual interpretation, are to be sought outside the human soul, Kepler's archetypes are within the soul. As he summarizes at the end of the discussion, the soul carries "not an image of the true pattern [*paradigma*], but the true pattern itself. . . . Thus finally the harmony itself becomes entirely soul, nay even God."[24]

[23] Letter to Herwart, April 9/10, 1599. Galileo later expressed the same principle: "That the Pythagoreans had the science of numbers in high esteem, and that Plato himself admired human understanding and thought that it partook of divinity, in that it understood the nature of numbers, I know very well, nor should I be far from being of the same opinion." de Santillana, *op. cit.*, p. 14. Descartes's remark, "You can substitute the mathematical order of nature for 'God' whenever I use the latter term" stems from the same source.

[24] For a discussion of Kepler's mathematical epistemology and its relation to neo-Platonism, see Max Steck, *Über das Wesen des Mathematischen und die*

This, then, is the final justification of Kepler's search for mathematical harmonies. The investigation of nature becomes an investigation into the thought of God, Whom we can apprehend through the language of mathematics. *Mundus est imago Dei corporea,* just as, on the other hand, *animus est imago Dei incorporea.* In the end, Kepler's unifying principle for the world of phenomena is not merely the concept of mechanical forces, but God, expressing Himself in mathematical laws.

Kepler's Two Deities

A final brief word may be in order concerning the psychological orientation of Kepler. Science, it must be remembered, was not Kepler's original destination. He was first a student of philosophy and theology at the University of Tübingen; only a few months before reaching the goal of church position, he suddenly—and reluctantly—found himself transferred by the University authorities to a teaching position in mathematics and astronomy at Graz. A year later, while already working on the *Mysterium Cosmographicum,* Kepler wrote: "I wanted to become a theologian; for a long time I was restless: Now, however, observe how through my effort God is being celebrated in astronomy."[25] And more than a few times in his later writings he referred to astronomers as priests of the Deity in the book of nature.

From his earliest writing to his last, Kepler maintained the direction and intensity of his religio-philosophical interest. His whole life was one of uncompromising piety; he was incessantly struggling to uphold his strong and often nonconformist convictions in religion as in science. Caught in the turmoil of the Counter-Reformation and the beginning of the Thirty Years' War, in the face of bitter difficulties and hardships, he never compromised on issues of belief. Expelled from communion in the Lutheran

---

mathematische Erkenntnis bei Kepler, Die Gestalt (Halle: Max Niemeyer, 1941), Volume V. The useful material is partly buried under nationalistic oratory. Another interesting source is Andreas Speiser, *Mathematische Denkweise,* Basel: Birkhäuser, 1945.
25 Letter to Mästlin, October 3, 1595.

Church for his unyielding individualism in religious matters, expelled from home and position at Graz for refusing to embrace Roman Catholicism, he could truly be believed when he wrote, "I take religion seriously, I do not play with it,"[26] or "In all science there is nothing which could prevent me from holding an opinion, nothing which could deter me from acknowledging openly an opinion of mine, except solely the authority of the Holy Bible, which is being twisted badly by many."[27]

But as his work shows us again and again, Kepler's soul bears a dual image on this subject too. For next to the Lutheran God, revealed to him directly in the words of the Bible, there stands the Pythagorean God, embodied in the immediacy of observable nature and in the mathematical harmonies of the solar system whose design Kepler himself had traced—a God "whom in the contemplation of the universe I can grasp, as it were, with my very hands."[28]

The expression is wonderfully apt: so intense was Kepler's vision that the abstract and concrete merged. Here we find the key to the enigma of Kepler, the explanation for the apparent complexity and disorder in his writings and commitments. In one brilliant image, Kepler saw the three basic themes or cosmological models superposed: *the universe as physical machine, the universe as mathematical harmony, and the universe as central theological order.* And this was the setting in which harmonies were interchangeable with forces, in which a theocentric conception of the universe led to specific results of crucial importance for the rise of modern physics.

[26] Letter to Herwart, December 16, 1598.
[27] Letter to Herwart, March 28, 1605. If one wonders how Kepler resolved the topical conflict concerning the authority of the scriptures *versus* the authority of scientific results, the same letter contains the answer: "I hold that we must look into the intentions of the men who were inspired by the Divine Spirit. Except in the first chapter of Genesis concerning the supernatural origin of all things, they never intended to inform men concerning natural things." This view, later associated with Galileo, is further developed in Kepler's eloquent introduction to the *Astronomia Nova.* The relevant excerpts were first translated by Thomas Salusbury, *Mathematical Collections* (London: 1661), Part I, pp. 461–467.
[28] Letter to Baron Strahlendorf, October 23, 1613.

QUESTIONS AND SUGGESTIONS

1. Study Holton's article to see how he uses historical source material. Then choose works written by a scientist who lived at least a century earlier than you and explain what the meaning of the works are, citing the text to support your argument. Explain the significance of the scientist. You will probably need to use primary and secondary sources; cite them in your bibliography.

2. Why would an audience of physicists be interested in Holton's article?

3. How do Kepler's science and his religious beliefs interrelate? Does science relate to religion in the contemporary world?

JAMES WATSON
AND FRANCIS CRICK

Molecular Structure of Nucleic Acids

JAMES WATSON and FRANCIS CRICK felt that they had "unlocked the secret of life" when they found the structure of the DNA molecule.

A Structure for Deoxyribose Nucleic Acid

We wish to suggest a structure for the salt of deoxyribose nucleic acid (D.N.A.). This structure has novel features which are of considerable biological interest.

A structure for nucleic acid has already been proposed by Pauling and Corey.[1] They kindly made their manuscript available to us in advance of publication. Their model consists of three intertwined chains, with the phosphates near the fibre axis, and the bases on the outside. In our opinion, this structure is unsatisfactory for two reasons: (1) We believe that the material which gives the X-ray diagrams is the salt, not the free acid. Without the acidic hydrogen atoms it is not clear what forces would hold the structure together, especially as the negatively charged phosphates near the axis will repeal each other. (2) Some of the van der Waals distances appear to be too small.

Another three-chain structure has also been suggested by Fraser (in the press). In his model the phosphates are on the outside and the bases on the inside, linked together by hydrogen bonds. This structure as described is rather ill-defined, and for this reason we shall not comment on it.

[1] Pauling, L., and Corey, R. B., *Nature*, 171, 346 (1953); *Proc. U.S. Nat. Acad. Sci.*, 39, 84 (1953).

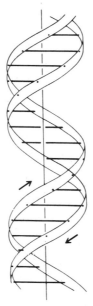

This figure is purely diagrammatic. The two ribbons symbolize the two phosphate-sugar chains, and the horizontal rods the pairs of bases holding the chains together. The vertical line marks the fibre axis.

We wish to put forward a radically different structure for the salt of deoxyribose nucleic acid. This structure has two helical chains each coiled round the same axis (see diagram). We have made the usual chemical assumptions, namely, that each chain consists of phosphate diester groups joining β-D-deoxyribofuranose residues with 3′,5′ linkages. The two chains (but not their bases) are related by a dyad perpendicular to the fibre axis. Both chains follow right-handed helices, but owing to the dyad the sequences of the atoms in the two chains run in opposite directions. Each chain loosely resembles Furberg's[2] model No. 1; that is, the bases are on the inside of the helix and the phosphates on the outside. The configuration of the sugar and the atoms near it is close to Furberg's "standard configuration," the sugar being roughly per-

2 Furberg, S., *Acta Chem. Scand.*, 6, 634 (1952).

pendicular to the attached base. There is a residue on each chain every 3·4 A. in the z-direction. We have assumed an angle of 36° between adjacent residues in the same chain, so that the structure repeats after 10 residues on each chain, that is, after 34 A. The distance of a phosphorus atom from the fibre axis is 10 A. As the phosphates are on the outside, cations have easy access to them.

The structure is an open one, and its water content is rather high. At lower water contents we would expect the bases to tilt so that the structure could become more compact.

The novel feature of the structure is the manner in which the two chains are held together by the purine and pyrimidine bases. The planes of the bases are perpendicular to the fibre axis. They are joined together in pairs, a single base from one chain being hydrogen-bonded to a single base from the other chain, so that the two lie side by side with identical z-co-ordinates. One of the pair must be a purine and the other a pyrimidine for bonding to occur. The hydrogen bonds are made as follows: purine position 1 to pyrimidine position 1; purine position 6 to pyrimidine position 6.

If it is assumed that the bases only occur in the structure in the most plausible tautomeric forms (that is, with the keto rather than the enol configurations) it is found that only specific pairs of bases can bond together. These pairs are: adenine (purine) with thymine (pyrimidine), and guanine (purine) with cytosine (pyrimidine).

In other words, if an adenine forms one member of a pair, on either chain, then on these assumptions the other member must be thymine; similarly for guanine and cytosine. The sequence of bases on a single chain does not appear to be restricted in any way. However, if only specific pairs of bases can be formed, it follows that if the sequence of bases on one chain is given, then the sequence on the other chain is automatically determined.

It has been found experimentally[3,4] that the ratio of the amounts of adenine to thymine, and the ratio of guanine to cytosine, are always very close to unity for deoxyribose nucleic acid.

It is probably impossible to build this structure with a ribose

[3] Chargaff, E., for references see Zamenhof, S., Brawerman, G., and Chargaff, E., *Biochim. et Biophys. Acta,* 9, 402 (1952).
[4] Wyatt, G. R., *J. Gen. Physiol.,* 36, 201 (1952).

sugar in place of the deoxyribose, as the extra oxygen atom would make too close a van der Waals contact.

The previously published X-ray data[5,6] on deoxyribose nucleic acid are insufficient for a rigorous test of our structure. So far as we can tell, it is roughly compatible with the experimental data, but it must be regarded as unproved until it has been checked against more exact results. Some of these are given in the following communications. We were not aware of the details of the results presented there when we devised our structure, which rests mainly though not entirely on published experimental data and stereochemical arguments.

It has not escaped our notice that the specific pairing we have postulated immediately suggests a possible copying mechanism for the genetic material.

Full details of the structure, including the conditions assumed in building it, together with a set of co-ordinates for the atoms, will be published elsewhere.

We are much indebted to Dr. Jerry Donohue for constant advice and criticism, especially on interatomic distances. We have also been stimulated by a knowledge of the general nature of the unpublished experimental results and ideas of Dr. M. H. F. Wilkins, Dr. R. E. Franklin and their co-workers at King's College, London. One of us (J. D. W.) has been aided by a fellowship from the National Foundation for Infantile Paralysis.

QUESTIONS AND SUGGESTIONS

1. In light of Watson's account of the work on DNA in *The Double Helix,* the assertion in the *Nature* article, "It has not escaped our notice that the specific pairing we have postulated immediately suggests a possible copying mechanism for the genetic material" is one of the great understatements of all time. What purpose did the statement serve? What was the overall purpose of the swift publication in *Nature*? Who was the audience for the article? How do this purpose and

5 Ashbury, W. T., *Symp. Soc. Exp. Biol.* 1, Nucleic Acid, 66 (Camb. Univ. Press, 1947).
6 Wilkins, M.H.F., and Randall, J. T., *Biochim. et Biophys. Acta,* 10, 192 (1953).

audience differ from those of the Charles Darwin article? Howard Ensign Evans?

2. What can you deduce from this article about the circumstances of Watson and Crick's discovery that is illuminated by the biographical account in the first section of the anthology? What does the article conceal?

3. What are the reasons for the order of presentation of the various parts of this article?

JENS FEDER

Invasion Percolation

JENS FEDER (b. 1939) is a professor of physics at the University of Oslo, where he also received his Ph.D. in 1970. He has been a visiting scientist with IBM's Zürich and Yorktown Heights, New York, research facilities and a visiting scientist with General Electric in 1978, 1979, and 1986. Feder's primary research is in statistical physics, and in 1972 he won the *Norsk Varekrigsforsikrings fonds* Prize for outstanding contributions in several areas of solid state physics. Feder is a fellow of the American Physical Society and a member of the Norwegian Academy of Science and Letters. He is the author of numerous articles published in a variety of specialized and generalist journals as well as the books *Fractals* (1988) and *Fractals in Physics* (1989) with Amnon Aharony. In 1988, one reviewer in *Nature* called the following chapter on invasion percolation (the invasion of porous media by fluids) "a clear discussion of this phenomenon and its relation to fractal concepts" and *Fractals* "the book to support a university course in the study of fractals." After reading this excerpt, do you agree with the reviewer?

Invasion percolation is a *dynamic* percolation process introduced by Wilkinson and Willemsen (1983), motivated by the study of the flow of two immiscible fluids in porous media (de Gennes and Guyon, 1978; Chandler et al., 1983). Consider the case in which oil is displaced by water in a porous medium. When the water is injected very slowly then the process takes place at very low capillary numbers Ca. . . . This implies that the capillary forces completely dominate the viscous forces, and therefore the dynamics of the process is determined on the pore level. In the limit of vanishing capillary numbers one may neglect any pressure drops both in the *invading fluid* (water) and in the *defending fluid* (oil). However, there is a pressure difference between the two fluids (the capillary pressure) given by

$$(p_{\text{invader}} - p_{\text{defender}}) = \frac{2\,\sigma\,\cos\theta}{r}$$

. . . Here σ is the interfacial tension between the two fluids, θ is the contact angle between the interface and the pore wall, and r is the radius of the pore at the position of the interface.

One often finds that water is the "wetting" fluid and oil the "non-wetting" fluid, i.e., the contact angle $\theta > 90°$, and the water will spontaneously invade the oil-filled porous medium unless the pressure in the water is kept below that of the defending oil. The important point to note is that the pressure difference depends on the local radius of the pore or pore neck where the interface lies. In a porous medium one must have variations in r (and possibly in the contact angle) and the interface must adjust to positions so that the above equation is satisfied everywhere. The capillary forces are strongest at the narrowest places in the medium. Thus if all the throats are smaller than all the pores, the water–oil interface moves quickly through the throats, but gets stuck entering the larger pores. It is consistent with both a simple theoretical model and experimental observations to represent this motion as a series of discrete jumps in which at each time step the water displaces oil from the smallest available pore.

Wilkinson and Willemsen (1983) proposed to simulate this process in an idealized meduim where the network of pores may be viewed as a regular lattice in which the sites and bonds of the lattice represent the pores and the throats respectively. Randomness of the medium is incorporated by assigning random numbers to the sites and bonds to represent the sizes of these pores and throats. Simulation of the process in a given realization of the lattice thus consists of following the motion of the water–oil interface as it advances through the smallest available pore, marking the pores filled with the invading fluid.

This model also applies in the case where a nonwetting fluid, say air, displaces a wetting fluid. In this case the pressure in the invading fluid is above that of the defending fluid, and the interface advances quickly through the large pores and gets stuck in the narrow throats connecting the pores. An illustration of the types of structures observed in this case is shown in Figure 1.

It is apparent in Figure 1 that the invading fluid *traps* regions of the defending fluid. As the invader advances it is possible for it to completely surround regions of the defending fluid, i.e., completely disconnect finite clusters of the defending fluid from the exit sites of the sample. This is one origin of the phenomenon of "residual oil," a great economic problem in the oil industry. Since

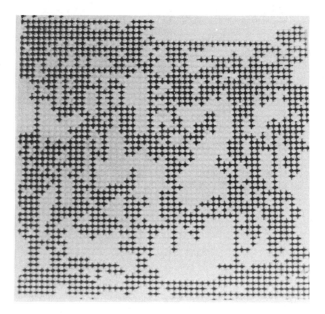

Figure 1. Invasion percolation of water (black) displacing air in a model consisting of a regular array of cylinders 2 mm in diameter and 0.7 mm high separating parallel plates. The water does not wet the model. The water enters at the upper left corner and exits at the lower right corner. The displacement is at Ca $\simeq 10^{-5}$ (Feder et al., 1986).

oil is incompressible, Wilkinson and Willemsen introduced the new rule that water cannot invade trapped regions of oil.

The algorithms describing invasion percolation are now simple to describe:

- Assign random numbers r in the range [0,1] to each site of an $L \times L$ lattice.
- Select sites of injection for the invading fluid and sites of extraction for the defending fluid.
- Identify the *growth sites* as the sites which belong to the defending fluid *and* are neighbors to the invading fluid.

- Advance the invading fluid to the growth site that has the lowest random number r.
- TRAPPING: Growth sites in regions completely surrounded by the invading fluid are not active and are eliminated from the list of growth sites.
- End the invasion process when the invading fluid reaches an exit site.

This model advances the invading fluid to new sites one by one, always selecting the possible growth site with the lowest random number associated with it. This is an algorithm that lets the invading cluster grow in a manner subject to *local* properties. The rule that a trapped region cannot be invaded introduces a nonlocal aspect into the model. The question whether or not a region is trapped cannot be answered locally, and involves a *global* search of the system.

It is interesting to compare the invasion percolation process without trapping to the ordinary percolation process. . . . In the ordinary percolation process one may grow percolation clusters as follows: The sites on an $L \times L$ lattice are assigned random numbers r in the range [0,1], and one places a seed on the lattice. Then for a given choice of the occupation probability p, $0 \leq p \leq 1$, the cluster grows by occupying all available sites with random numbers $r \leq p$. The growth of the percolation cluster stops when no more such numbers are found on the boundary (perimeter) of the cluster. . . . Only if the seed site happens to lie on the incipient percolating cluster at p_c, or on the percolating cluster for $p > p_c$, will the percolation cluster grow to a size that spans the lattice. By contrast, in invasion percolation the cluster grows by always selecting the smallest random number, no matter how large. However, once a large number r_0 has been chosen, it is not necessarily true that subsequently every number $r \geq r_0$ will be chosen—smaller numbers will in general become available at the interface, and will thus be chosen. The cluster shown in Figure 2 is grown by this process until it reaches the edge of the system.

Wilkinson and Willemsen (1983) simulated the invasion process on lattices of size $L \times 2L$, injecting the invader on the left-hand edge and stopping the simulation at the *breakthrough* point, when

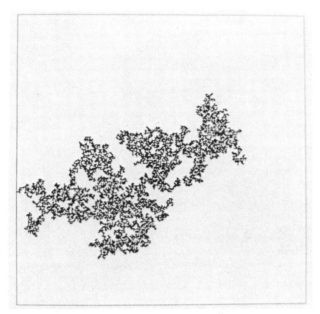

Figure 2. An invasion percolation (without trapping) cluster grown from the central site on a 300 × 300 lattice until it reaches one of the edges of the lattice. The cluster contains 7656 sites.

the invader reached the right-hand edge. We illustrate this geometry in Figure 3.

Naturally, in the finite geometry the invader will gradually fill the entire lattice if the invasion process is continued. Wilkinson and Willemsen found that the number of sites $M(L)$ in the central $L \times L$ portion of the lattice at breakthrough increases with the size of the lattice as follows:

$$M(L) = A\, L^{D_{\text{inv}}}, \text{ with } D_{\text{inv}} \simeq 1.89$$

This equation is analogous to the equation, $M(L) \sim \overline{A} L^D$, $D = 91/48 = 1.895$, and the fractal dimension of invasion percolation without trapping, D_{inv}, is found to equal the fractal dimension of the incipient percolation cluster at p_c. There is now considerable evidence that invasion percolation in fact is in the same universality class as ordinary percolation (Dias and Wilkinson, 1986).

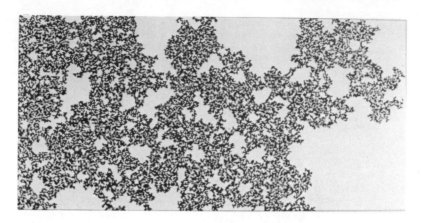

Figure 3.   *Invasion percolation (without trapping) on a 100 × 200 quadratic lattice. The invader (black) enters from sites on the left-hand edge and the defender escapes through the right-hand edge. At "breakthrough" the invader just reaches the right-hand edge.*

Trapping changes the invasion percolation quite drastically in two dimensions. In Figure 4 we show an invasion cluster grown by the process that includes trapping. Comparing the invasion percolation cluster in Figure 3 to the invasion percolation cluster with trapping shown in Figure 4 one sees that the trapping rule leads to clusters with much larger holes in them. This is reflected in how the number of sites, $M(L)$, that belong to the central part of an $L \times 2L$ lattice, with injection from one side, scales with the size of the lattice,

$$M(L) = A \, L^{D_{\text{trap}}}, \text{ with } D_{\text{trap}} \simeq 1.82$$

This result was obtained by Wilkinson and Willemsen (1983). Lenormand and Zarcone (1985a) investigated *experimentally* the invasion of air at a very slow rate into a two-dimensional network of 250,000 ducts of random widths on a square grid that was filled by glycerol (see Figure 5). Note that they continued the experiment after breakthrough. By placing a semipermeable membrane at the right-hand edge of the sample cell they could prevent the invading air from escaping, and the invasion process was terminated when all the remaining defending fluid was trapped. They found that

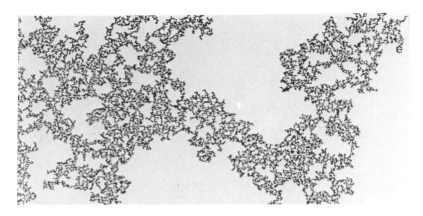

*Figure 4. Invasion percolation with trapping on a 100 × 200 qua-
dratic lattice. The invader (black) enters from sites on the left-hand
edge and the defender escapes through the right-hand edge. At
"breakthrough" the invader just reaches the right-hand edge.*

the number of ducts filled by the invading fluid counted inside
boxes of side L followed the above equation with $1.80 < D_{\text{trap}} <
1.83$, consistent with the numerical simulations.

We have experimentally generated invasion percolation clusters
with the trapping rule by injecting air at the center of a two-dimensional circular model of a porous medium consisting of a layer of
glass spheres placed at random and sandwiched between two plates.
The resulting cluster, shown in Figure 6, was found to have a
fractal dimension of $D = 1.84 \pm 0.04$, also consistent with the expected result.

We conclude that the invasion percolation process with trapping
generates fractal structures, which have a fractal dimension that is
lower than the fractal dimension both of invasion percolation without trapping and of ordinary percolation. The experimental results
for immiscible fluid displacement in two-dimensional porous media
at very low capillary numbers are consistent with the process of
invasion percolation with trapping introduced by Wilkinson and
Willemsen (1983).

However, in three dimensions the situation is altogether different. Consider ordinary percolation in the simple cubic (s.c.) lattice,

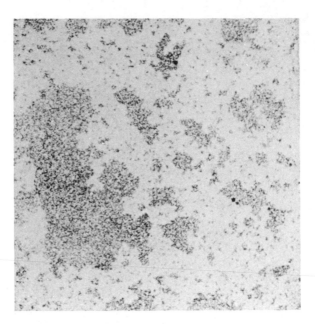

Figure 5. Displacement of the wetting fluid (black) by the non-wetting fluid (white) injected on the left-hand side of the network. On the right-hand side, a semipermeable membrane prevents the nonwetting fluid from flowing outside (Lenormand and Zarcone, 1985a).

where each lattice site has six neighbors. The percolation threshold for this geometry is $p_c(\text{s.c.}) \simeq 0.3117$, and the fractal dimension of the incipient percolation cluster at p_c is $D \simeq 2.5$. The important point to note is that in this case there is a range $p_c \leqslant p \leqslant (1 - p_c)$ of occupation probabilities for which *both* the occupied sites *and* the empty sites percolate and form spanning clusters. By contrast in two-dimensional percolation on the square lattice one finds that either the occupied sites or the empty sites percolate and there is *no* range for which both percolate. Interestingly, percolation on the triangular lattice is a borderline case since $p_c = 0.5$ is the simultaneous threshold both for the occupied sites and for the empty sites.

The qualitative difference between two and three dimensions

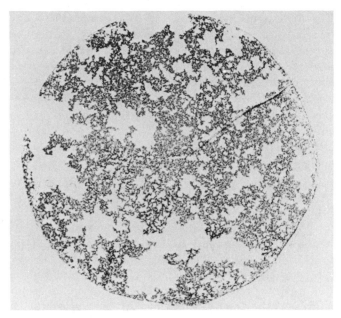

*Figure 6.   Air (black) displacing glycerol (white) at very low capil-
lary numbers Ca ≃ 10⁻⁵, in a two-dimensional model consisting
of a layer of randomly packed glass spheres 1 mm in diameter. The
fractal dimension of the cluster of invading of air is* D ≃ *1.84
(Måløy et al., 1987b).*

extends to invasion percolation. Wilkinson and Willemsen (1983)
find $D \simeq 2.52$ for the invading cluster at breakthrough, both with
and without the trapping rule. The existence of a range of occupation probabilities for which both the defending fluid and the
invading fluid percolate makes trapping much less effective. Most
of the sites are still filled by the defending fluid when the invading
fluid percolates the three-dimensional sample from one face to the
other. When the invasion process is continued one finds that the
defending fluid is trapped at $p = (1 - p_c)$. Wilkinson and Willemsen found that at this point the number of sites in the invading
cluster increased as L^2, so that it in fact represented a *finite fraction* $\simeq 0.66$ of the sites of the simple cubic lattice of size $L \times L \times L$.
The invading cluster is therefore *not* fractal at the limit where the

defending fluid is trapped. Dias and Wilkinson (1986) have discussed a related model, *percolation with trapping*, which includes the trapping rule but ignores the invasion part of the problem. They analyze the size distribution of the trapped regions and they give strong evidence for the conclusion that the critical behavior of invasion percolation with trapping belongs to the same universality class as ordinary percolation in three dimensions.

Experimentally it is difficult to realize the three-dimensional invasion percolation process. Clément et al. (1985) injected nonwetting Woods metal into consolidated crushed glass very slowly from the bottom and analyzed both vertical and horizontal cuts of the cylindrical sample. They concluded that the horizontal cuts, perpendicular to the flow direction, show that the Woods metal had invaded the porous medium in a self-similar manner, and the fractal dimension of the distribution of the metal in the cut was found to be $\simeq 1.65$. This is somewhat above the fractal dimension, 1.50, expected for a cross-section of the incipient percolation cluster. However, gravity effects cannot be neglected and clearly influence the results, and the fractal dimension of the horizontal cuts depends on the level at which they are taken.

QUESTIONS AND SUGGESTIONS

1. This article relies heavily on the previous research of Wilkinson and Willemsen. Locate their essay "Invasion Percolation: A New Form of Percolation Theory" in *Journal of Physics* 16 (1983): 3365–3376; prepare an executive summary of their essay and then a report on the bearing of their findings on Feder's conclusions.

2. Benoit B. Mandelbrot is considered "The Father of Fractals" and, since 1967, he has been writing and publishing articles on this mathematical subject. Prepare an annotated bibliography of some of Mandelbrot's writings on fractals or an annotated bibliography on the subject of fractals.

3. Write a summary and critique of "Invasion Percolation," paying special attention to Feder's language and use of visuals. What role do the equations serve?

4. Write a report on the history of fractals for a college-level liberal arts student.

5. Locate one of the following books for a book report: Feder, *Fractals;* Heinz-Otto Peitgen and Dietmar Saupe, *The Science of Fractals* (Janson, 1989); or L. Pietronero and E. Tossatti, eds., *Fractals in Physics* (Dutton, 1986).